ATM & SONET BASICS

George Dobrowski & Don Grise

ATM & SONET Basics

First Printing: January 2000

Published by
Intertec Publishing Corporation
9800 Metcalf Avenue
Overland Park, KS 66212-2215

Library of Congress Catalog Card Number: 99-067680
ISBN 0-87288-718-9

ACKNOWLEDGEMENTS

We would like to acknowledge the contribution of Candice Murray to this book. In addition, we would like to thank Bill Rubin and Laura Lippold of Telcordia Technologies who provided valuable editorial review and comment.

This book is dedicated to the memory of Harry E. Young.

TABLE OF CONTENTS

1 INTRODUCTION

In today's competitive computing and telecommunications markets, users are adapting and expecting ever-increasing speeds and functionality from computers and communications systems. Video displays, program and media interaction, multi-user conferences, access to information from Web locations and shared media are common, whereas a few years ago these were "pipedreams." Computing power at users' fingertips is exploding by doubling every two years or faster. Users are more dispersed when they need to communicate.

All of these market forces drive the "need for speed" between computing, display and interactive devices. New regulations and competition are changing the traditional boundaries between the telecommunications, cable, data, and broadcast industries that will enable consolidated networks to carry all types of traffic. Digitization of information makes this possible. A bit of information does not know if it is Broadcast, Cable, Telephony, Electronic Print, Computer Network information, etc. thereby making content free from the type of transport mechanism. Bits can be transmitted over any media, readily stored, processed anywhere, and customer devices can easily manipulate digital bits. Some recent demonstrations of digitization blurring all boundaries include electric utilities offering telecommunications services, broadcasters offering telecommunications services, telephone industry offering broadcast services. However, what will be the underlying technology basis or combination of technologies for the infrastructure(s), packet routing or switching? This question represents an unprecedented challenge. Also, debates continue between the advocates of a more random-based best effort packet addressing and routing scheme such as Internet Protocol (IP) versus a more structured switched approach providing higher capacities and guaranteed quality levels.

Actually, a hybrid of these approaches appears to be emerging. For terminal access to network communications, IP is emerging as the most effective. For an intranetwork communications scheme and for the backbone infrastructure, a more structured, high-volume, guaranteed performance (such as bandwidth guarantees) approach to communications is needed.

Packet switching provides a very flexible transport mechanism, but its network processing functions are complex. Circuit switching provides a very inflexible transport mechanism, but its network processing functions are very simple. For the majority of other switching technologies, their network flexibility is defined by the complexity of the processing functions. Asynchronous Transfer Mode (ATM) is a hybrid incorporating both circuit and packet switching techniques into one technology. ATM is a special case technology with fixed size packets or cells. It provides a flexible transport network solution with simple network processing functions.

The routing versus switching debate makes it appear that there is competition between ATM and Internet (IP) technologies. In reality the two technologies are both fundamental to future network architectures, sometimes referred to as Next Generation Network (NGN). NGNs will leverage off the strengths of each, sometimes converging and sometimes with complementing capabilities.

The speed at which networks shift towards NGN will significantly impact the existing embedded circuit switched digital network. The effect of residential and small business Internet packet access on the embedded circuit switched telephony base is just the beginning. As all forms of communication go to all packet or Variable Bit Rate (VBR) digital techniques, systems will need to be in place to keep pace and provide not only the necessary performance, but flexibility in meeting still emerging needs.

This book provides a basic understanding of the ATM principles and how it

Technology	Application	Speed
Circuit Switching	Voice	64 Kbps to 1.544 Mbps
Ethernet	Data	10 Mbps, 100 Mbps, 1 Gigabit
Token Ring	Data	4 Mbps, 16 Mbps, or 20 Mbps
FDDI	Data, Video, Voice	100 Mbps, may also provide 1 Gigabit in the future
X.25 Packet Switching	Data	56 Kbps to 1.544 Mbps
Frame Relay	Data	56 Kbps, 1.544/2.048 Mbps, or 44.736 Mbps[1]
SMDS	Data	64 Kbps to 34/44.736 Mbps
ATM	Data, Video, Voice	Any speed with granularity from 1 ATM Cell to multi-Gigabits bounded by state of the art for silicon, switch architectures, and physical media.
IP Routing	Data with Voice and Video Emerging	Any speed and granularity up to the protocol processing and routing limits

Summary of Switching Technologies
Table 1-1

[1] Higher rate Frame Relay is generally supported over ATM

provides a more structured signaling and control method for high-speed, flexible and efficient intra and inter network communications. Along with ATM, this book will highlight elements of the Synchronous Optical Network (SONET) which is a physical layer high-speed fiber optic multiplexing arrangement often used with ATM applications in the network.

In North America, the American National Standards Institute - Telecommunications (ANSI/T1) has adopted ATM and Synchronous Optical NETwork (SONET) as the basis for the Broadband Integrated Digital Services Network (B-ISDN) transmission and switching technologies.

Also, the International Telecommunication Union - Telecommunications Sector (ITU-T) has adopted ATM and Synchronous Digital Hierarchy (SDH) which is based on SONET as the B-ISDN transmission and switching technology. SDH uses similar digital hierarchies to SONET with small differences between the two formats which will be discussed later. ATM and SONET/SDH technologies are the basis for the optical physical interfaces for the User-Network Interface (UNI) and Network - Network Interface (NNI), which provide high-speed data transfer and supports various communication modes. See Table 1-1.

The intent of this book is to familiarize the reader with the concept of ATM and SONET. It also provides a path to a fundamental understanding of ATM and SONET and basic elements of their interworking. The layout of this book is designed to look at ATM and SONET separately as well as how they can work together. At the end of the book, examples of ATM applications are discussed.

PART 1

ATM Basics

2 PRINCIPLES OF ATM

Asynchronous Transfer Mode (ATM) is an integrated multiplexing and packet switching technique originally designed for high-speed (155.52 Mbps and above) optical physical layer SONET/SDH broadband networking. It uses fixed data units or "cells" to transfer information from the source to the destination. The ATM layer defines the cell structure and how ATM cells flow through the logical connections in a network. A cell consists of an information field (cell payload), which is transported transparently, and a header. A label field inside each cell header is used to define and recognize individual communications that are multiplexed over the same shared media. In this respect, ATM resembles conventional packet transfer protocols. Also, similar to packet switching techniques, ATM can provide communication with a bit rate that is individually tailored to the actual need of the user application. This includes time-variant bit rates. The term "asynchronous" refers to the fact that cells allocated to the same connection may exhibit an irregular recurrence pattern because they are filled before transmission, according to the actual user application characteristic.

Building on these fundamental characteristics, ATM is designed to be a general purpose service-independent, connection-oriented transfer mode that can be used for a wide range of services and applied to LAN, public and private network technologies. ATM handles connection-oriented traffic either directly over ATM (referred to as Cell Relay or native ATM) or through adaptation layers for service-specific support. It also handles connectionless traffic using an adaptation layer. ATM connections may operate at either a Constant Bit Rate (CBR) or Variable Bit Rate (VBR). The label field of each ATM cell header sent into the network contains address information that is used to establish a Virtual Channel (VC). All cells associated with a connection are transferred in sequence. ATM provides either Permanent Virtual Channel (PVCs) or Switched Virtual Connections (SVCs). The transfer capacity of each ATM connection may be assigned on demand (through signaling) depending on resource availability.

The ATM layer provides ATM connection switching (cell relaying), multiplexing and demultiplexing, in-band layer management, generic flow control and some basic traffic control. It supports multiple grades of service based on loss and/or delay priorities. The ATM layer preserves the cell sequence integrity on an ATM Layer connection. In addition to segmenting and reassembling information frames, ATM attributes enable a scaleable, high performance, seamless technology able to simultaneously support multimedia applications, voice, video and many forms of data. The ATM layer introduces network management and operations mechanisms for fault detection and isolation for improved reliability. It defines and sets the industry standard for traffic management and Quality of Service (QoS) parameters.

ATM's appeal as a unifying technology for a global communications fabric spans from Local Area Networks (LANs), campus networks, entire business enterprise networks to Wide Area Networks (WANs). While ATM's impact is pervasive, ranging from semiconductor innovations to protocol standards, hardware and software design, shrink-wrapped applications, it is transparent to the end user. ATM's ability to mimic legacy technologies and interwork with embedded systems is essential for easy user introduction and migration.

2.1 ATM CELL STRUCTURE

The ATM cell consists of a 5-octet header and a 48-octet payload for user data and is illustrated in Figure 2-1. The payload can carry any type of user information, including voice, various types of data and video. The use of the payload field and the ATM Adaptation Layer (AAL) Types are discussed in Section 7. The ATM layer works in conjunction with the Physical layer to establish cell delineation or identification of where a cell starts based on what type of physical media signaling technique is used. This will be discussed in more detail in Section 5, but essentially ATM has been adapted to numerous physical layers with the objective of alleviating the need to re-wire. For example, in the case

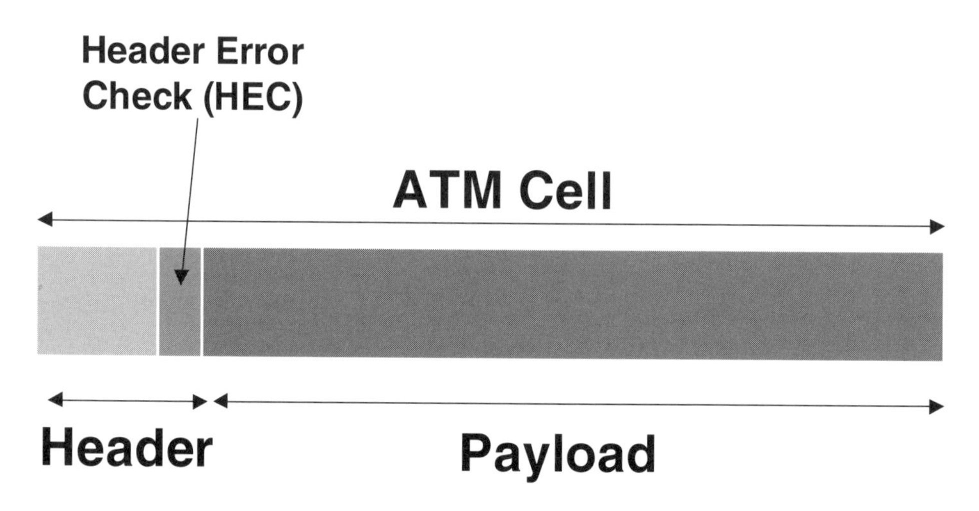

ATM Cell Structure
Fig. 2-1

where Fiber Distributed Data Interface (FDDI) multi-mode fiber and silicon chips employ either 4B/5B or 8B/10B coding, selected code symbols are used to delineate ATM cells. With 4B/5B coding, 5-bit symbols are encoded into 4-bit data to assure timing recovery, synchronization and start of cell. In the case of 8B/10B coding, 10-bit symbols are encoded in 8-bit data. The receiver locks onto the 5-octet header "blocks" which satisfy the Header Error Check (HEC) calculation performed in real time in the receiver chip. The HEC is coded in a manner in which empty cells will not appear to be headers. The cell header provides a number of additional functions such as identifying the destination, cell type, priority, traffic and network management.

The fields defined in the 5-byte header are Generic Flow Control (GFC), Virtual Path Identifier (VPI), Virtual Channel Identifier (VCI), Payload Type (PT), Cell Loss Priority (CLP) and Header Error Control (HEC). Two different encoding schemes for the cell header are adopted according to the type of interface being considered, either the User Network Interface (UNI) or the Network Network Interface (NNI). The payload, however, is unchanged. The UNI is the interface between the end customer equipment and the network switch. The NNI is the interface used between switches or between networks. See Figure 2-2, Figure 2-3, and Figure 2-4.

Generic Flow Control (GFC): The GFC is a 4-bit field intended to provide a mechanism to assist in controlling the multiplexing of traffic on a UNI that is shared by multiple users. The function is similar to IEEE 802.6 Distributed Queue Dual Bus (DQDB) standard with a shared access medium (such as a Local Area Network multipoint shared physical media) but use of the field was never defined as part of the standard. Currently it is being considered as a means of controlling multiplexer contention for shared resources.

Virtual Connection Identifier (VPI/VCI): The Virtual Path Identifier (VPI) and Virtual Channel Identifier (VCI) are label fields used to identify the destination

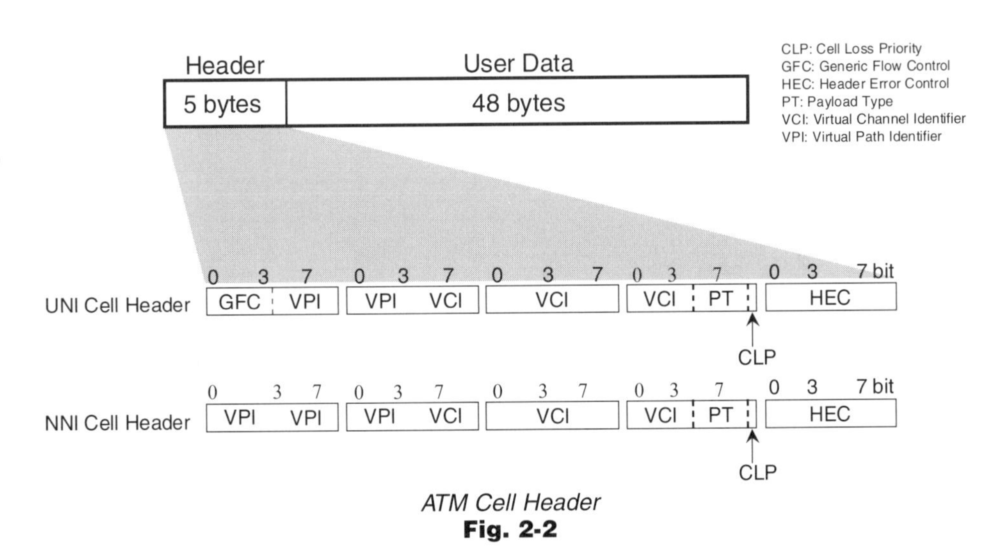

ATM Cell Header
Fig. 2-2

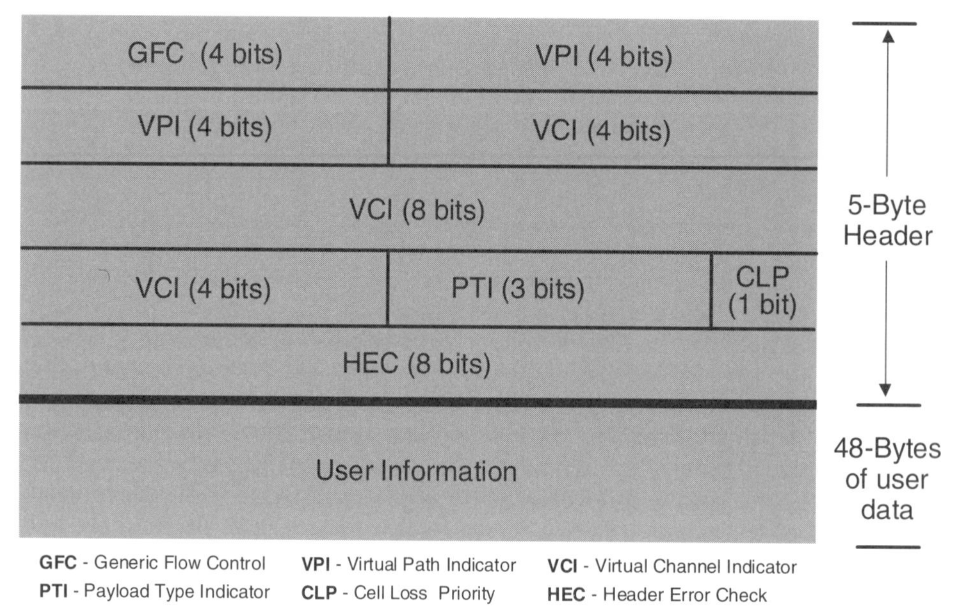

UNI (User Network Interface) Cell Header Format
Fig. 2-3

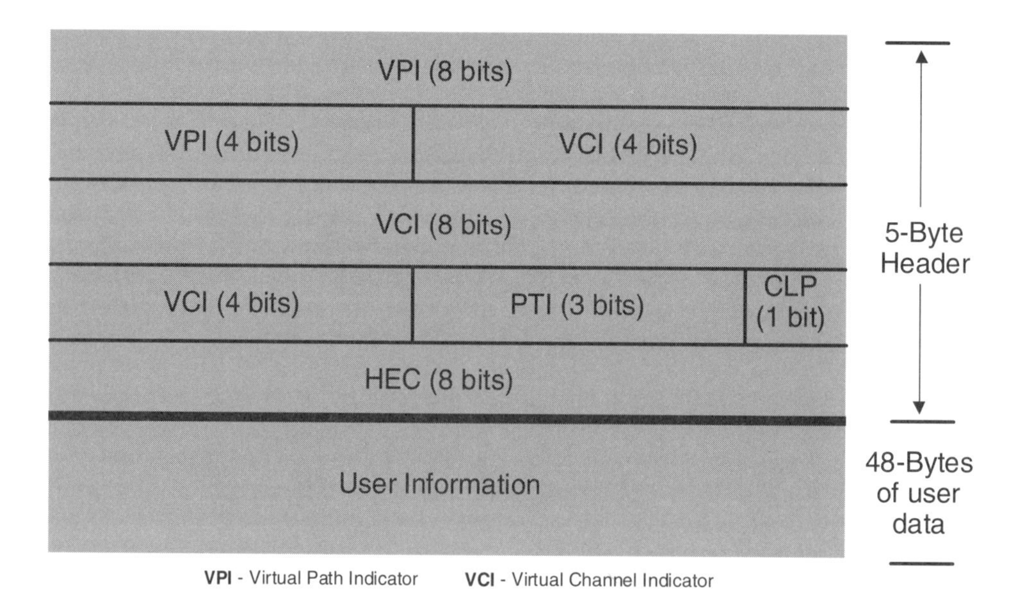

NNI (Network Node Interface) Cell Header Format
Fig. 2-4

of a connection. VPI and VCI are used for ATM cell switching and routing. The VPI/VCI only have local significance applying to a given interface connection and do not extend through the switch or network on an end-to-end basis. Use of VPIs/VCIs is described further in this Section.

The UNI VPI/VCI field consists of 24 bits divided into two subfields: 8 bits for VPI, and 16 bits for VCI, see Figure 2-3.

The NNI VPI/VCI field consists of 28 bits: 12 bits for VPI, and 16 bits for VCI. For the NNI, the GFC function is not provided, and the additional bits are allocated for VPI use, see Figure 2-4.

Payload Type (PT): The 3-bit PT field is used to provide an indication of whether a cell contains user-upper layer information or a cell carrying layer management information in the payload. The PT field is from bit 4 to 6 of the fourth octet in the header. Table 2-1 depicts the PT field coding details. The user data and Operations, Administration & Maintenance (OA&M) messages travel the exact same channel thereby providing a powerful fault detection and isolation capability not available with other technologies. The OA&M cells are able to verify continuity of a connection without having to interject a test signal with the user information and to enable isolating the location of a failure with loop-back tests. For example, OA&M cells assist in determining whether a connection failure is on the ingress access link from the originating customer, on a section within the network, on the egress/destination access link or within the customers equipment attached at either end.

Cell Loss Priority (CLP): This 1-bit field is used to indicate the relative cell loss priority. In the event of network congestion, the CLP assists the network in selectively discarding cells and minimizing the degradation of service that may be perceived if the network encounters congestion. If the CLP is set, (meaning

				PT Value
User Data Type 0 indicates ATM cell is not last cell of Protocol Data Unit (PDU). Type 1 indicates last cell of PDU	No congestion		Type = 0 Type = 1	000 001
	Congested		Type = 0 Type = 1	010 011
OA&M Payload contains Layer Management	F5 Flow		OAM F5 Cell Segment OAM F5 End-to-End Cell	100 101
	Traffic		Resource Management Reserved Future Use	110 111

Payload Type Field
Table 2-1

it has a value of 1), the cell is subject to discard, depending on network conditions. If the CLP is not set (the value is 0), the cell has a higher priority and should not be discarded.

Header Error Control (HEC): The 8-bit HEC field applies to the 5-octet ATM cell header. HEC code provides single-bit error correction and detects of multiple-bit errors over the cell header. Since the cell header VPI/VCI field is used to route and switch cells to the proper destination, it is important that errors are detected before delivery of information to the incorrect user occurs or an incorrect action is taken in the ATM layer. While HEC is part of the cell header, it is used by the Transmission Convergence (TC) part of the physical layer to provide error control and cell delineation functions. The HEC is processed by the physical layer. The TC part of the Physical Layer generates the HEC on cell header transmissions and uses the HEC to determine if the received cell header has any errors. The HEC normally operates in the correction mode where is corrects single-bit errors. When a correction is performed, the HEC goes into the error-detection mode. In the detection mode, it detects multiple-bit errors and discards the cell when additional errors are detected. If no errors are detected for a period of time, it resumes correction mode.

The characteristics of ATM's fixed-size cells and error checks on the ATM cell header reduces switching delay (queuing delay). ATM is a connection oriented packet switching technology where a connection is established through the network before user information flows. This is in contrast to connectionless packet switching where a packet is sent/routed from one point to the next without regard to whether a complete path exists to the destination point or whether the resources are available. The packet is forwarded from one point to the next just like a letter in an addressed envelope. The ability to provide Quality of Service (QoS) and dynamic bandwidth allocation on a per connection basis provides flexibility for ATM traffic and congestion control enabling support of a multi-service (voice, data, video) communications environment. The key characteristics are:

- High speed
- Low delays
- Fixed-cell length
- Connection-oriented
- Dynamic bandwidth allocation and management
- Scalability

2.2 CHOICE OF ATM CELL PAYLOAD AND REASON FOR FIXED-SIZE CELLS

The size of the ATM cell was hotly debated in the ITU-T when developing the original 13 Recommendations that standardized the principles of ATM. Over a period of time the debate centered on 32-octet versus 64-octet payload size and was polarized along global regional vested interests. The decision on the 48-octet payload size was a compromise essentially based on "equal pain" for all. The 48-octet size meant that no vendor or country would have a competitive advantage in getting products to market. Everyone had to change their prototype designs. The 5-octet header was chosen after reviewing 3-octet, 6-octet

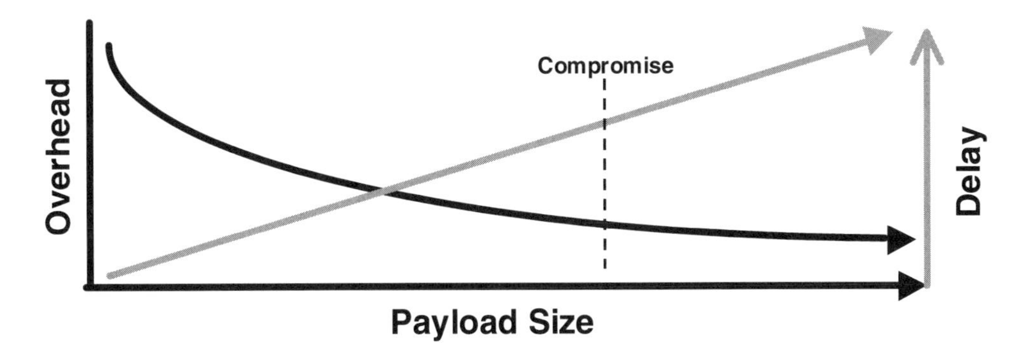

- Large cells are great for data, poor for voice
- Small cells are great for voice, poor for data
- 5+48 cell yields 8% overhead for data with 6 ms packetization delay for voice

Why Use Such Tiny Cells?
Fig. 2-5

and 8-octet header proposals. The final choice was influenced by agreements on the requirements and functionality the header had to provide.

Why use fixed-size cells? As shown in Figure 2-5, the cell size is a tradeoff between the amount of overhead each cell or packet requires and the delay experienced by packets queued behind other waiting packets to be transmitted. In larger cells, the overhead represents a smaller percentage of the packet information. However, the larger the cell, the greater the delay experienced by other cell/packet applications queued up waiting for transmission while the previous cell/packet is sent. The use of variable size packets also requires complex buffer allocation and variable processing rates in handling the packets. This is especially true when different priorities or quality of service are introduced. Fixed size cells allow pipelining of cell processing functions. Buffer allocation is simplified when all units of information are of the same size, enabling deterministic processing for real time dependent applications such as voice and video.

2.3 ATM MULTIPLEXING

Unlike Time Division Multiplexing (TDM) in traditional telephony and narrowband ISDN, ATM does not assign time slots to a specific channel. It dynamically allocates ATM cells into the next available time slot. When two cells arrive at the same time, the slot decision is made according to the cell priority, Quality of Service (QoS), and connection type. Simple illustrations are shown in Figure 2-6 and Figure 2-7. In Figure 2-6, the first example shows that, during the same six time slots, TDM is only able to send two packets of data. This is because TDM requires that all channels get their share in a fixed sequence, even if there is no information to send. In the second example, ATM is able to

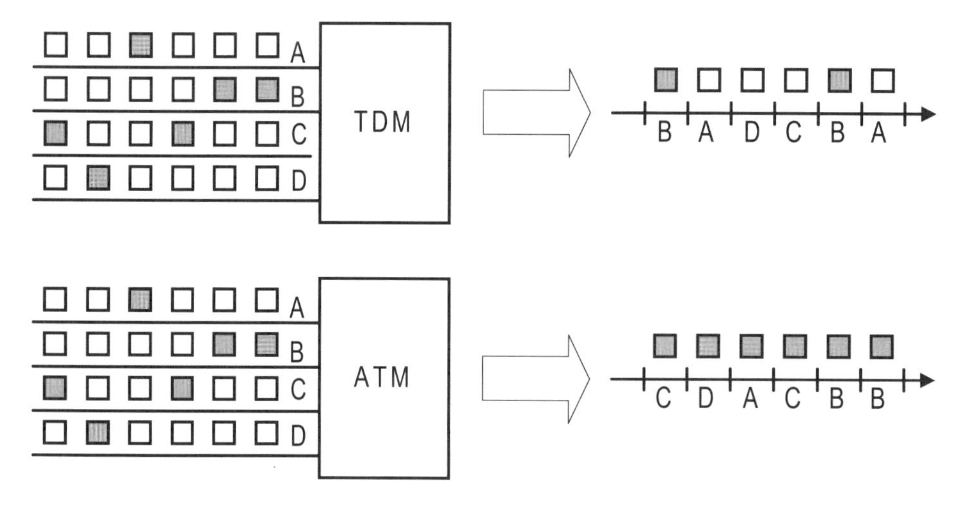

ATM vs. TDM Multiplexing
Fig. 2-6

utilize all six time slots by allowing the channel, that has data to send, to utilize the idle time slots.

TDM, sometimes also called Synchronous Transfer Mode (STM), has a rigid structure with a predefined amount of bandwidth. In ATM networks, the multiplexing and switching of cells is independent of the actual application. In principle, the same piece of equipment can handle a low bit-rate connection as well as a high bit-rate connection and regardless of whether the information is a

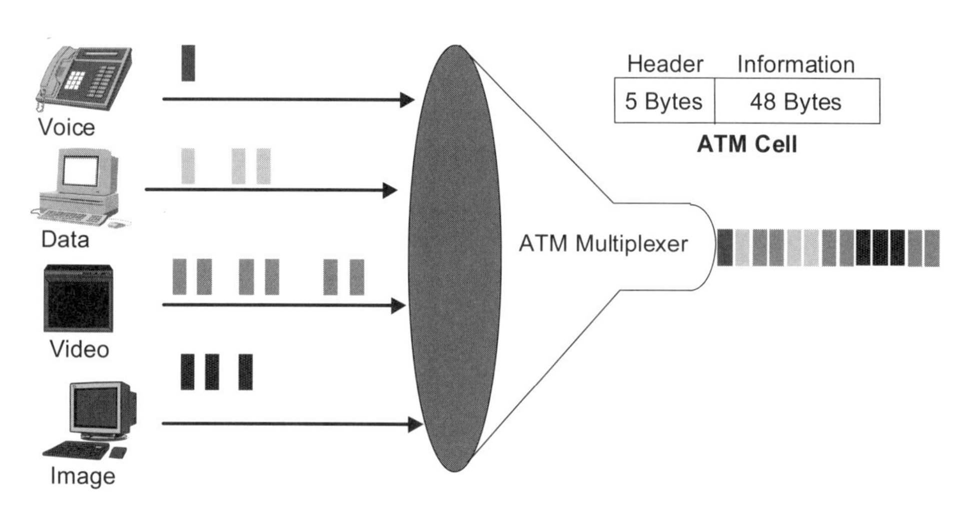

ATM is a multiplexing and switching technique based on fixed-size packets called cells

Asynchronous Transfer Mode (ATM) Multiplexing
Fig. 2-7

stream or a burst. ATM provides the ability to integrate multiple media over the same facility, to operate at any speed, to use both packet switching and circuit switching techniques, and to support continuous and bursty applications. ATM is a standardized international technology enabling a single technology platform for all applications.

2.4 ATM CONNECTION MODES AND CONNECTION TYPES

The types of ATM connection modes are characterized as Constant Bit Rate (CBR), where the information flow maintains a steady stream, and Variable Bit Rate (VBR) connections. With VBR connections, the flow of information is not periodic and is less predictable and bursty in nature. ATM connection types include point-to-point, point-to-multipoint, multipoint-to-point and multipoint-to-multipoint.

- Point-to-point connections connect two ATM end systems. The connections can be uni-directional or bi-directional.
- Point-to-multipoint connections connect the source end system (the root node) to multiple destination end systems (the leaves). The connections are uni-directional from the root out to the leaves.
- Multipoint-to-point connections have the same configuration as point-to-multipoint. However, the transmission direction is uni-directional from the leaves to the root.
- Multipoint-to-multipoint connections consists of multiple overlaid point-to-multipoint connections.

2.5 ATM NETWORKING BASICS: VIRTUAL PATH AND VIRTUAL CHANNEL CONCEPT

ATM introduces the basic networking concepts of the Virtual Path (VP), and the Virtual Channel (VC). These concepts form the basic building blocks of ATM networks.

Figure 2-8 graphically illustrates the relationship between the physical transmission path, and VPs and VCs. The transmission path contains one or more virtual paths, and each virtual path contains one or more virtual channels. Multiple virtual channels can be trunked into a single virtual path. Switching can be performed on either a transmission path at the physical layer (such as protection switching of SONET/SDH facilities), virtual path or virtual channel basis.

The ATM cell structure and the ATM layer defines how information carried in cells flows through the logical VP and VC connections. A VP is a bundle or collection of Virtual Channel connections made through an ATM network. Each VP and VC can be individually established in one of two ways, either permanently established or dynamically set up for the duration of time necessary to transmit the user information. Virtual Channel connections and Virtual Path connections that are set up permanently are referred to as Permanent Virtual Connections (PVCs) and Permanent Virtual Paths (PVPs) respectively. VC and VP connections that are dynamically established use the access signaling control protocol to communicate the characteristics of the connection desired from the user to the

- Thousands of Virtual Channels (VCs) can be carried in Virtual Paths (VPs)
- Hundreds/ thousands of VPs can be carried in a Physical Link
- Virtual Channel Identifier (VCI) is contained in ATM cell header
- Virtual Path Identifier (VPI) is contained in ATM cell header
- VCI, VPI are read at network elements (such as ATM switches) and customer equipment which process ATM layer information
- Bandwidth can be flexibly assigned on a per VC basis
- End-to-end VC Connection (VCC) is concatenation of VCs between customer and network equipment elements

Virtual Path (VP) and Virtual Channel (VC) Concept
Fig. 2-8

network. Depending on the version of the signaling protocol, the characteristics of the connection (such as traffic/bandwidth and QoS parameters) can be negotiated during set up. If the requested attributes are not available, they may be re-negotiated[2]. Dynamically established VC and VP connections are switched through the network based on the ATM cell header connection identifier field or Virtual Path Identifier (VPI) and Virtual Channel Identifier (VCI), and are referred to as Switched Virtual Connections (SVCs) and Switched Virtual Paths (SVPs).

2.6 ATM LAYER PROCESSING

The high performance processing of ATM cells is enabled by the small, fixed-size cell format. The ATM cell header is only 5 octets long and requires minimal processing as cells arrive and flow through equipment. In fact, the ATM principles are simple. Fully utilizing the capabilities that ATM provides involves development of additional functions that build on top of the ATM layer. These additional functions include ATM adaptation layers, signaling, traffic management, and network management capabilities. These are discussed in later sections.

As each ATM cell arrives, the ATM layer is first checked for errors by processing the header checksum field, referred to as HEC. If no errors are present, the cell header is checked to determine if the cell contains user information or whether the cell is for internal network use only [such as signaling, network management or maintenance check/test (loop around) information]. If the cell contains user information, then the Virtual Channel Identifier (VCI) and Virtual Path

[2] Bellcore GR-1111-CORE for User/Network Access Signaling, Issue 2 Bellcore, October 1996, and ATM *Forum UNI Signaling version 4.0* define procedures for dynamic negotiation. ITU-T Q.2931 Recommendation has not yet been completed to support negotiation capability.

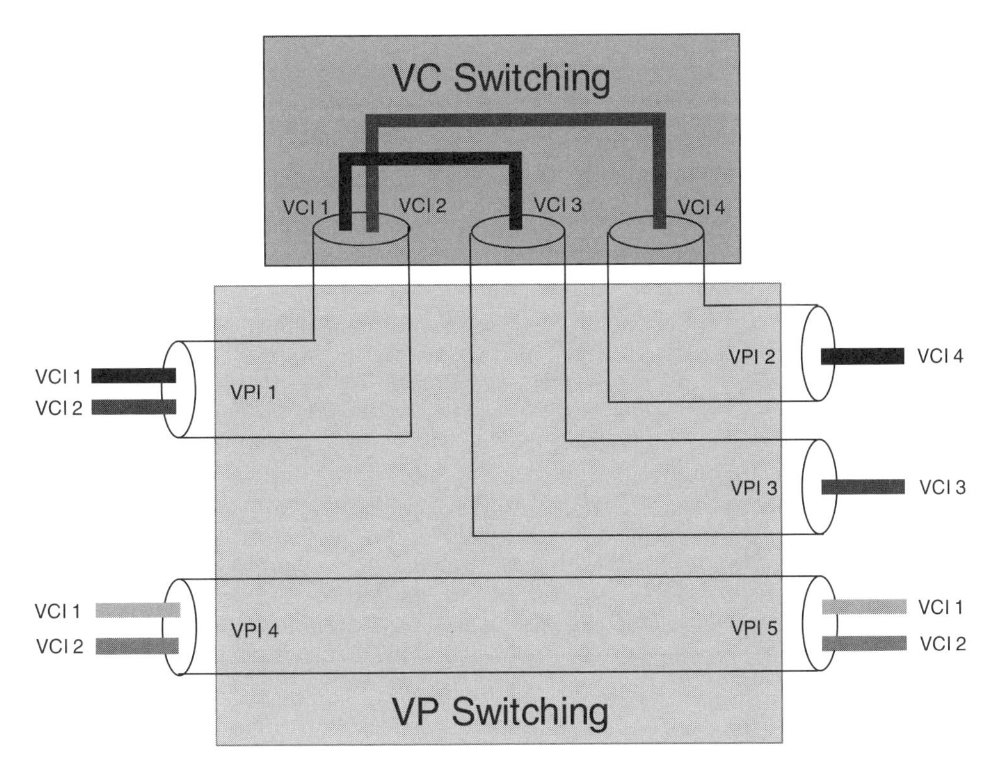

Switching ATM Virtual Channel
Fig. 2-9

Identifier (VPI) are looked up in the routing table of the switching equipment to determine to which outgoing facility the cell should be switched and routed. Because the outgoing connection involves a different physical port, and it is likely that the incoming channel address may already be used on the outgoing port for another connection, the switch replaces or maps the input VPI/VCI value to the new VPI/VCI values prior to sending them through the ATM switching fabric. See Figure 2-9. For example, VCI 1 on VPI 1 is mapped to outgoing VPI 3 and assigned to VCI 3 contained in that VP. VCI 2 within incoming VPI 1 is destined for a different location and is mapped to outgoing VPI 2 and assigned to VCI 4 within VPI 2.

ATM switching is fast. In performing all of the above ATM functions after the initial connection has been set up, the maximum ATM cell switching delay requirement is defined as 150 microseconds.[3] Switch delay refers to the time it takes to transmit a cell from its ingress through the switch fabric and then to the egress switch port. Switch vendors generally exceed this worst case switch delay requirement, making switch delay a point of vendor product differentiation.

[3] Bellcore GR-1111-CORE, *Broadband Switching System (BSS) Generic Requirements*, Issue 1, 1994, with Revisions Through 6, January 1999 (Bellcore).

To simplify processing and increase performance, multiple VC connections can be bundled into one VP. The VP can be switched and routed in whole, without having to examine each individual VC connection. The expectation is to simplify bundling of all traffic going to the same destination such as another switch. Traffic management and policing functions are performed only on the aggregated VP level. Consequently, the quality of service characteristics of the bundled VC connections must be that of the most stringent VC carried within that bundled path because only the VPI part of the cell header is processed in the case of VP switching. Further, CBR connection types can not be mixed with VBR connection types, nor can VBR "best effort" connections be mixed/bundled with VBR connections that have QoS guarantees.

The ATM switch can also add/change certain quality of service information in the ATM cell such as the cell loss priority bit and provide current congestion information in resource management cells before transmission out of the switch. If you contrast this with TCP/IP-based Internet and Intranet technology, the IP packet sizes vary considerably, and packet headers contain much more information that must be processed at each link/hop by router processors throughout the network. This requires more computing power and buffer space within every route server in the network than does ATM. ATM switches perform equivalent functions when the packet connection is initially established. Once set up, information "flows" through the switch without re-computation at every hop in the network.

2.7 VP/VC SWITCHING

VP and VC switching of logical links involves mapping the incoming VPI/VCI cell header "address" with its outgoing VPI/VCI address, as determined by the switch routing table. The cell leaves the switch with a new VPI/VCI header setting. When the end-to-end connection is established, the network assigns the VCI/VPI values and sets up the routing tables at each node along the link. All cells of this transmission travel on the same link to the destination. At the end of the communications session, either end user can release the connection, which causes the corresponding routing table entries for the SVC to be deleted at each intermediate node.

ATM network links are set up in two ways: Permanent Virtual Connections (PVC) and Switched Virtual Connections (SVC). PVC requires some manual configuration to set up the switch routing table via the use of an operations support system. This type of connection does not require dynamic call control or processing capabilities, but bandwidth management is still necessary. SVCs are configured automatically through the use of signaling protocols. SVCs connections require signaling for call/connection establishment and for tear down and capacity management. PVC is not a preferred method for a large-scale network because the manual configuration can be difficult to manage with thousands or millions of subscribers.

In VP switching, VCs are multiplexed into VPs, and travel through switches with the VCI address field unchanged although the VPI address may be changed from node to node. This can be viewed as a special case of VP/VC

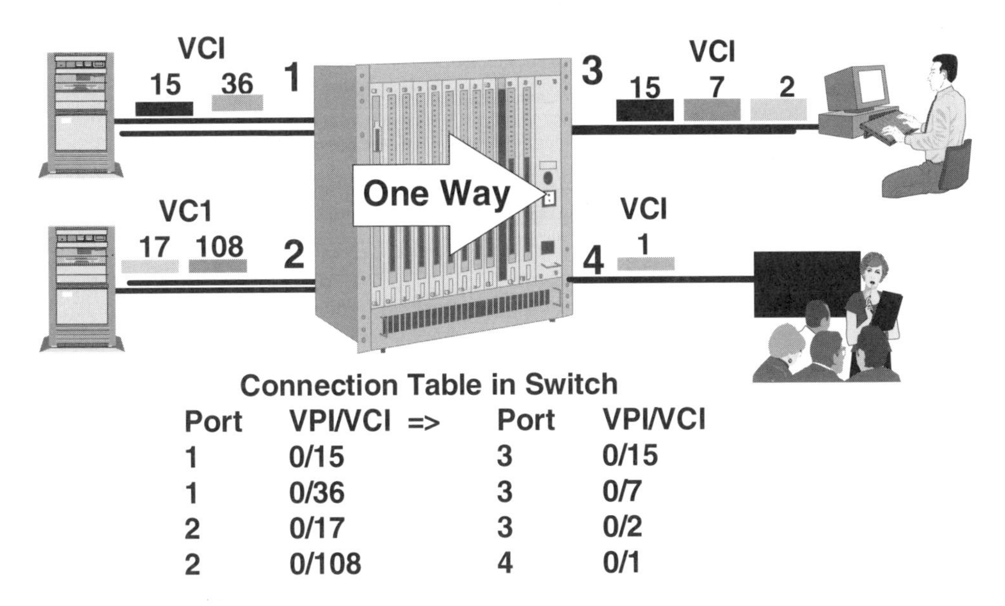

Connection Table in Switch

Port	VPI/VCI =>	Port	VPI/VCI
1	0/15	3	0/15
1	0/36	3	0/7
2	0/17	3	0/2
2	0/108	4	0/1

VP/VC Switching
Fig. 2-10

switching. It reduces the routing table size and network control complexity. In VP switching, the VPs are classified as those terminated at the switch and those not terminated at the switch. The latter method is also used for VP cross-connects. If the VP is terminated at the switch, a predefined VCI used for control signaling determines the new VPI of the output port.

3 ADVANTAGES OF ATM

3.1 SUPPORT FOR VOICE, VIDEO AND DATA APPLICATIONS

ATM technology provides service integration of voice, video and data applications. Voice and video applications are real-time or near real-time type applications that tolerate very low delay before service degrades. On the other hand, data applications are a burst-type transmission tolerant to switching and routing delay. ATM networks have the flexibility to carry voice, video and data on the same consolidated multi-service network infrastructure while independently meeting the quality of service needs of each user. Besides point-to-point connectivity, ATM also supports multipoint connectivity for applications such as video conferencing, distributed multimedia and two-way video. ATM offers a number of advantages that will enable ATM networks to become the underlying infrastructure. This will enable users to take advantage of new computing and communications power.

3.2 EFFICIENT USE OF NETWORK RESOURCES

ATM allocates bandwidth dynamically (known as bandwidth-on-demand) according to user needs and resources available within the network. The ATM mechanism reserves network resources for user applications based on bandwidth and QoS requirements. This enables real-time dependent applications to be guaranteed resources when needed to adhere to performance requirements, and at the same time it allows unused resources to be shared by those applications that require only "best effort" service.

Certain applications such as audio and real-time video demand a deterministic level of service. ATM's Virtual Connection (VC)-oriented methodology and ability to negotiate QoS parameters ensure that multimedia application requirements of low latency and low cell delay variation can be met. Further, the VC capability ensures that one user's traffic does not degrade the services of another VC user on the same facilities.

3.3 FLEXIBILITY AND SCALABILITY

ATM is not restricted to certain speeds and distance

limitations. ATM was initially developed for optical fiber physical media and high-speed carrier networks, but ATM essentially applies to any physical media and includes wireless, copper twisted pair, coax, and various types of fiber (single mode and multi-mode fiber, glass and plastic fiber) operating over a wide range of speeds. ATM has been adapted to the transmission hierarchies around the world and to media used for local area networks. No re-wiring is necessary on the part of users or carriers/operators. ATM also supports packet communication techniques and traditional circuit switched techniques in one platform.

A shift in the industry is taking place. All applications are going to a packet or Variable Bit Rate (VBR) approach for voice and video as well as existing data applications. ATM-based products have anticipated this shift as well as the rate of data traffic growth with the Internet Protocol (IP). Since ATM technology is agnostic to the types of traffic, it is able to adapt well to this changing mix of traffic and provides both flexibility and the ability to "future proof" backbone infrastructure.

The ATM principles of Traffic Management (TM) and Quality of Service (QoS), work and scale equally well in the Local Area Network (LAN), campus networks, Metropolitan Area Networks (MAN), and Wide Area Networks (WAN). Scalability refers to ATM's ability to grow in terms of physical size (local and wide areas) and speed as well as in the number of users it can support without having one user's traffic/application interfere with that of another user. ATM works on an end-to-end basis. This allows a uniform method of transport, independent of the physical media and bandwidth or speed.

3.4 INTEGRATED SERVICES

ATM was designed to carry voice, data, image and video simultaneously over a common/shared infrastructure. The QoS guarantees bandwidth management via a VC-oriented cell-based operation, which enables the deployment of a single multi-service network where all traffic types can be transported. This eliminates the need for parallel and technology-specific overlay networks, thereby reducing operational and maintenance costs. The cost savings include those associated with having to keep spare inventory of parts, specialized technology-specific staff training and support, and network complexity. These factors strongly influence the true Total Cost of Ownership (TCO) and Return On Investment (ROI).

3.5 TRANSPARENCY TO EXISTING APPLICATIONS

ATM has the unique ability to emulate nearly any protocol. This is important because a tremendous amount of embedded equipment and applications exist. These applications will be around for a long time and no one is going to throw out their existing equipment and re-write application software so that they can use ATM. ATM provides new ways and alternatives for interconnecting and internetworking LANs. At the application software level, ATM is transparent. Circuit Emulation Service (CES) allows ATM to emulate existing private line (T1/E1 and T3/E3) transmission interfaces, supporting interconnection of embedded equipment such as PBXs, access multiplexers and pair gain systems over ATM without changes. The disadvantage of this approach, however, is that users and applications do not fully utilize the capabilities and performance that ATM offers.

3.6 INTERNETWORKING WITH EXISTING LANS AND WANS

ATM will serve as the underlying technology unifying existing local area and wide area network services offerings. ATM's ability to essentially mimic any protocol eases the introduction and migration toward ATM based solutions. The objective is to allow ATM to be introduced on a selective basis when and where needed. Consequently, a number of specifications have been developed, such as LAN Emulation (LANE) of Ethernet and Token Ring technologies (discussed further in Part 3, Applications and Standards) and Circuit Emulation Service for T1/T3 and E1/E3 circuits. In the wide area, Frame Relay, Narrowband ISDN (N-ISDN), and Switched Multimegabit Data Service (SMDS) interworking specifications are available. Perhaps the most essential is the set of specifications for IP over ATM which includes not only LAN, but Multi Protocols Over ATM (MPOA), and Private Network Node Interface (PNNI) discussed later. In addition, various RFC (Request For Comment) documents have been developed or are being developed by the Internet Engineering Task Force (IETF) for IP over ATM. These specifications enable a smooth migration and seamless integration with existing network applications and allow the introduction of ATM, when and where needed, without wholesale replacement of embedded equipment and software.

This is equally important to enterprise networks and service provider/ operator networks. Today, the average corporation in the United States has six (6) different networks[4]. These networks need to be upgraded at different rates. ATM is the only technology that can interconnect and interoperate with the disparate technologies used today (Ethernet, Token Ring, X.25, SNA, voice, fax, message storage and retrieval, and PBXs). ATM can be introduced into that part of the enterprise network needing higher performance transparent to the applications. It paves the way for migration and consolidation of other segments of the enterprise network onto the ATM platform. Service providers/operators need to upgrade their backbone infrastructure to meet growing data traffic demands while continuing to support existing applications such as voice.

[4] Gartner Group 1998 report..

4 ATM PROTOCOL REFERENCE MODEL (PRM)

4.1 ASYNCHRONOUS TRANSFER MODE (ATM)-BASED BROADBAND INTEGRATED SERVICES DIGITAL NETWORK (B-ISDN)

A number of forces have been driving towards broadband networks. Technologies that allow more information to be delivered to the user inevitability drive out old technologies. These technologies provide higher-speed backbones, new computer communications techniques, the doubling of processor power approximately every 18 months and include new switching technologies. Customer needs for higher-bandwidth applications in the business environment and residential services, such as Internet access and entertainment, which have widely different traffic and cost characteristics, are other concerns. Finally, competition and regulatory changes require flexible and scaleable solutions. These factors led to the realization and conception of Broadband ISDN (B-ISDN).

4.1.1 B-ISDN Principles

The emerging technologies of high-speed multiplexing, switching and optical transmission systems foreshadow the realization of Integrated Services Digital Networks (ISDN) with broadband capabilities. The appropriate application of these technologies can provide users and service providers with an enormous information transfer capacity that can be flexibly drawn upon to meet existing and future service needs. Because switching and transmission requirements of emerging applications cannot be known precisely, it is crucial that the capabilities of B-ISDN be flexible.

To understand B-ISDN, one must go back to the original ISDN standards. In 1984, the CCITT[5] adopted a series of recommendations dealing with integrated services over digital networks. The CCITT stated that "ISDN network ... provides end-to-end digital connectivity to support a wide range of services, including voice and non-voice services, to which users have access by a limited set of standard multi-purpose user-network interfaces." [1]

---

[5] CCITT (Consultative Committee for International Telegraph and Telephone), now known as the International Telecommunications Union-Telecommunications (ITU-T) for public telephony networks, and ITU-Radio (ITU-R) for radio/satellite related standards.

The digitized ISDN network is characterized by its two interfaces, Basic Rate Interface (BRI) access consisting of two 64Kbps channels (frequently referred to as B channels) and a 16Kbps D-channel for signaling and data. The other interface is the Primary Rate Interface (PRI). PRI provides a channelized interface rate of 1.544Mbps (generally structured as 23B + D@64Kbps) for T1 transmission hierarchy or 2.048Mbps (generally structured as 30B + D@64Kbps) for the E1 transmission hierarchy. Other structures are possible and consist of certain multiples of 64Kbps channels.

The important concept ISDN introduced was the ability to support multiple connections and different media over the same facilities at the same time. The channelized approach was applied and extended for Broadband-ISDN with the addition of Virtual Channels, Virtual Paths, and logical connections.

It was soon realized that ISDN was too limiting. In the 1980s, all information including voice, video, and image with good resolution began to migrate from analog to digital coding thereby driving up the need for bandwidth and faster switching and transport. Local area data networks (LANs) and wide area data networks drove packet switching technology and need for more capacity. Specific needs were met initially by dedicated technologies and networks. Further, considering the on going costs associated with the deployment and maintenance of parallel service-specific networks operation, maintenance, provisioning, etc., the total cost of ownership of dedicated overlay networks, when aggregated together, represents a significant cost to public carriers/service providers. In addition, dedicated networks require several distinct and separate subscriber access lines/interfaces, increasing costs to both the user and service providers. These factors drove the global telecommunications industry toward a single solution. After much technical debate, the (ATM) technique was selected.

Furthermore, a number of the signaling, service control, network management protocols and standards developed for ISDN were extensible. Subsequently, ATM broadband standards and industry implementation agreements have been developed leveraging existing standards. This is particularly important from the perspective of easing the transition from existing networks and applications, and it minimizes interworking functions between the existing equipment and ATM broadband network implementations.

4.1.2 What is B-ISDN?

B-ISDN is both a protocol model and an architecture. The objective was to extend the original ISDN concept of a single network. A key element of service integration on a single network is the provision of a wide variety of services to a broad spectrum of users. These users utilize a limited set of connection types and multipurpose user-to-network interfaces. Some of the users' needs taken into consideration [2] include the following:

- The need to provide flexibility to handle emerging demand for broadband services (candidate services described in ITU-T Recommendation I.211, Broadband Service Aspects, [3] for both the user and operator)
- The need for the availability of high-speed transmission, switching and signal processing technologies
- The need for improved data and image processing capabilities to users and

service providers

- The need to apply software advances in computer and telecommunications
- The need to integrate interactive and distribution services, and circuit and packet transfer modes into a universal broadband network

The goal in B-ISDN development was to define a protocol reference model that would be able to flexdibly support divergent application requirements.

4.2 B-ISDN PROTOCOL REFERENCE MODEL

The B-ISDN Protocol Reference Model (PRM) [4] was developed as a common framework to facilitate the development of B-ISDN protocols and readily identify critical protocol architecture issues. It models the interconnection and exchange of information in a B-ISDN environment. The model was developed using a layered communication architecture similar to the one developed by the International Standards Organization (ISO) [5].

Before delving into the complete B-ISDN PRM, a list of the basic items that need to be described for a complete protocol specification is given. In general, the specification of a protocol layer should include the following items [6]:

- A general description of the purpose of the layer and the services it provides.
- An exact specification of the services that the layer provides to the upper layer and the services that it expects to receive from the lower layer.
- The structure and relationship of the layer in terms of entities.
- The description of the interactions between the entities, including the informal operation of the entities and the types and formats of messages exchanged between these entities.

4.2.1 Description of the ATM Protocol Reference Model (PRM)

The PRM is introduced in two steps.

- First, the reader is introduced to a simple one-dimensional layer model
- Secondly, a three-dimensional layered model is described.

The one-dimensional ATM Protocol Reference Model consists of four layers: the application layer, the ATM Adaptation Layer (AAL), the ATM Layer and the Physical Layer. Some layers are further divided into sublayers. See Figure 4-1. The layers illustrated in the figure do not have a one-to-one mapping relationship with the Open Systems Interconnection (OSI) seven-layer protocol reference model[6] which defines transport network functions. For example, the AAL is comparable to selected combined functions of OSI layer 4 (transport control), layer 5 (session control) and layer 7 (application control). The ATM layer is comparable to the combined but selected functions of OSI layer 2 (data link control) and layer 3 (network control).

The reader may think of the layers of the ATM protocol model as envelopes, one inside another. The application layer of ATM protocol model carries user/application information such as voice, video and data. The application layer is wrapped inside the AAL which defines the 48 octet payload of user data of the ATM cell. The AAL is wrapped inside the ATM layer. The ATM layer function is to

[6] The protocol layers of OSI seven-layer protocol model are application control, presentation control, session control, transport control, network control, datalink control, physical control.

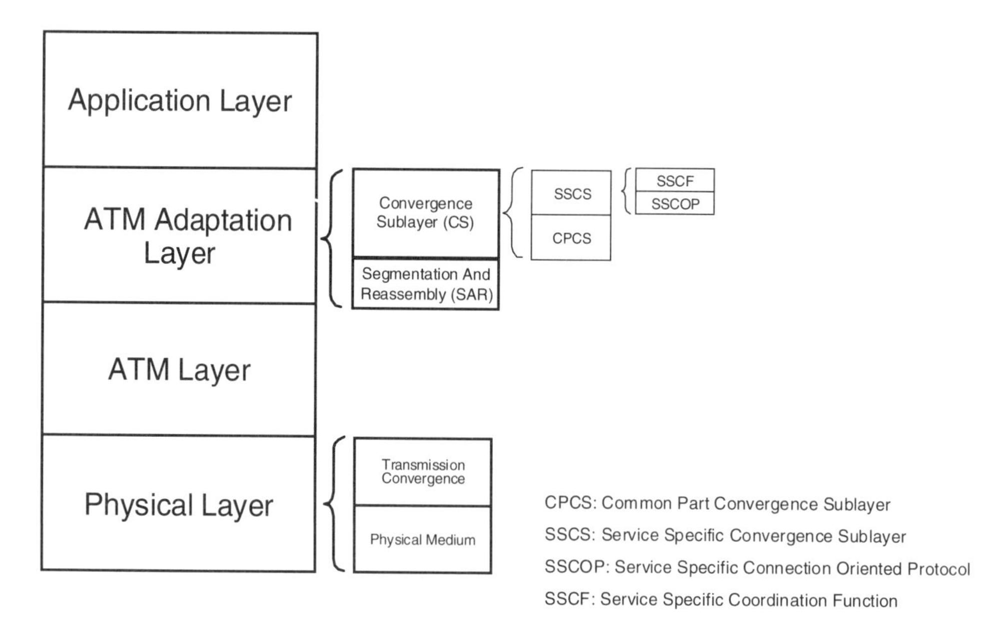

ATM Protocol Reference
Fig. 4-1

add and process the information of the 5 octets that constitutes the ATM cell header. It translates the routing information of VPI and VCI, checks header error, establishes cell priority at the switch, assists in traffic control, and provides Operation, Administration and Maintenance (OA&M) functions. The last layer of the ATM protocol is the physical layer which is associated with the ATM transport system.

The full B-ISDN PRM is based on the framework defined in ITU-T Recommendations I.121 [2] and I.321 [4]. While it uses similar layering concepts, the model had to be extended to meet the needs of B-ISDN to support functions such as signaling. This led to the concept of separated planes for the segregation of user information, control, and management functions.

The B-ISDN PRM is shown is Figure 4-2. This model contains the three structural elements named: User Plane, Control Plane, and Management Plane.

The User Plane, with its layered structure, provides for the transfer of user application information.

The Control Plane, also with a layered structure, handles the call and connection control functions, including signaling necessary to establish and release calls and connections, negotiation and allocation of network resources, etc. [7, 8, 9]. In-depth treatment of signaling can be found in the referenced documents.

The Management Plane provides for the management application functions. It contains a mechanism for information interchange between the User Plane and Control Plane processes. Functions related to the management aspects

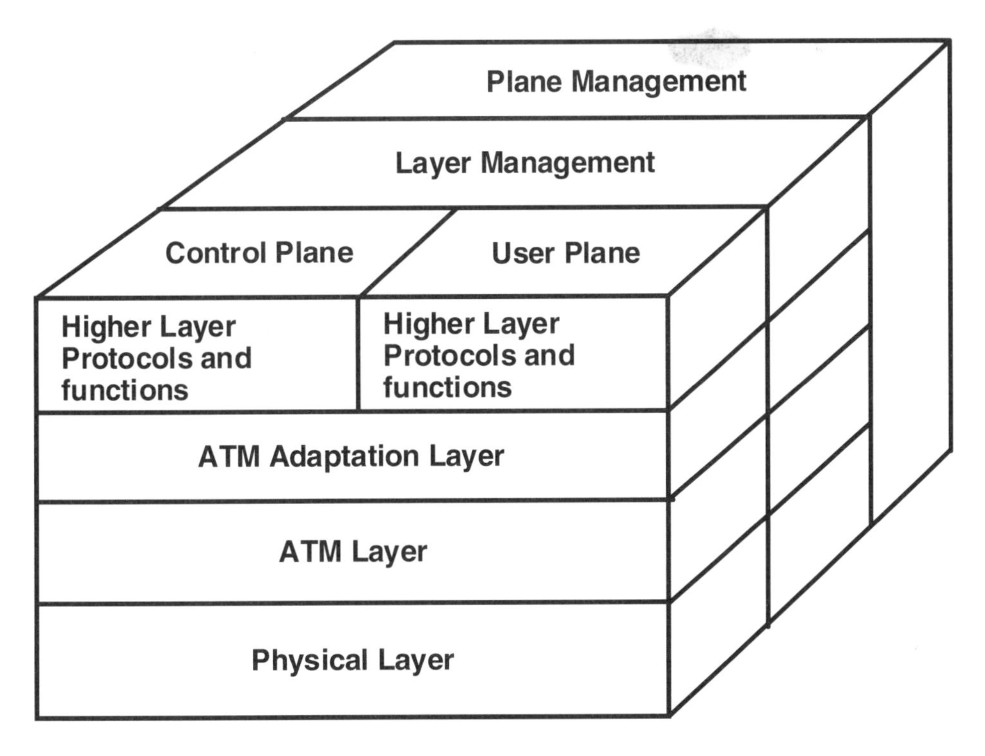

B-ISDN Protocol Reference Model
Fig. 4-2

include coordination of local operations across layers in establishing network connections, monitoring of established connections for failures, and responding to status queries to support network supervision.

The Management Plane has two sections: Layer Management and Plane Management. The Layer Management performs the management functions specific to a layer, such as layer-specific operation and maintenance (OA&M) tests/functions, and interacts with the layer entities and the Plane Management entities. The Plane Management section performs the management functions related to the system as a whole, the coordination between all the planes and the various Layer Management entities (LMEs). The Plane Management is not layered.

The ATM and physical layers are common to both the User Plane and the Control Plane. The functions of the Physical Layer are grouped into two sublayers: the Physical Medium Dependent (PMD) Sublayer, and the Transmission Convergence (TC) Sublayer. The PMD deals with the bit transmission over the physical medium of choice. While B-ISDN concepts initially assumed that the PMD would be based on SONET/SDH, the ATM principles apply to nearly any physical medium. Since the initial standards were developed, the ATM Forum[7] has adapted ATM to more than 20 physical interfaces. Some of the physical

[7] The ATM Forum is a research consortium formed in 1991 to speed the development and deployment of interoperable ATM products and services.

layer interfaces provide users a choice of physical media that best meets their needs, while other physical interfaces adapt ATM to the different transmission hierarchies used by service providers around the world. The TC Sublayer deals with the transmission framing and OA&M functions, as well as delineation of PHY data units such as identifying where an ATM cell starts. The ATM Layer [10 & 11] provides for the transparent and sequential transfer of fixed-size data units between a source and the associated destination(s) with an agreed Quality of Service (QoS). The PHY and ATM layers are service-independent, meaning that it only includes functions which are applicable or common to all services.

The ATM Adaptation Layer (AAL) performs the necessary functions to adapt the services provided by the ATM Layer to the services required by the different service users. Therefore, the AAL functions are service dependent, and several AALs with different protocol characteristics have been standardized [12]. Some of the functions of the AAL entities are targeted to support services that require Constant Bit Rates (CBR) and timing relationships such as circuit emulation. Other AAL functions address the support of Variable Bit Rate (VBR) bursty data traffic such as IP [13 & 14], and Switched Multmegabit Data Service (SMDS)[15] based connectionless data services or connection-oriented data services, such as Frame Relay [16 & 17], as well as the signaling control channel. Due to the extreme variability of the requested services to be supported, five different AALs protocols have been developed. Furthermore, the use of an AAL is optional. A user can operate without an AAL. This is referred to as the "null AAL," native ATM or Cell Relay (CR) service.

To keep things in perspective, the following sections provide more details on the Physical layer, ATM Adaptation layer, and the user, control and management plane functions to effectively utilize ATM.

5 PHYSICAL LAYER

The physical layer of the ATM Protocol Reference Model consists of the Transmission Convergence sublayer and the Physical Medium sublayer, Figure 5-1. The Transmission Convergence sublayer converts the cell stream into transportable bits. Its function includes cell rate decoupling, HEC sequence generation and verification, cell delineation, transmission frame adaptation, and transmission frame generation and recovery. The Physical Medium sublayer handles the bit timing function. When using SONET/SDH as an ATM transport system, the Physical layer is functionally divided into 3 sublayers. The functions of these three sublayers are discussed in detail in Part 2.

The Physical Media Dependent (PMD) sublayer is the lowest sublayer in the ATM Protocol Reference Model. It provides for the actual clocking of bit transmission over the physical medium and includes the physical medium itself. The ATM physical medium can be electrical, optical or radio/wireless. Some examples of physical medium are Unshielded Twisted Pair (UTP), Radio Frequency (RF)/satellite, and fiber optic. Using UTP provides the advantage that it is hard to break physically and can be low cost. The common UTP cabling used for ATM are Category 3 and Category 5. Using RF transmission for ATM networks provides for mobility. Using satellites to establish the ATM link provides the advantage of connecting to points that are not line-of-sight. Together, these media provide the infrastructure of tactical networks. For example, an Army troop can quickly establish

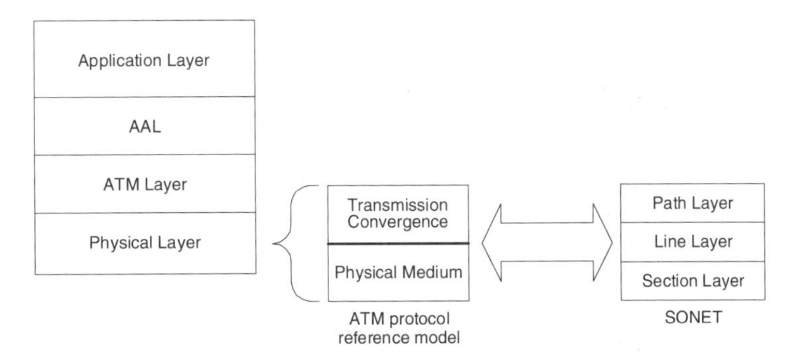

Two Ways of Looking at the Physical Layer
Fig. 5-1

its ATM network in the battlefield and tear it down before moving to a different location. Using fiber optics provides low signal loss, and it is the media used in SONET/SDH. The common fiber optics used for ATM include OC-3, OC-12 and OC-48. Higher rate transmission at OC-192 (10 Gigabit) and Wave Division Multiplexing (WDM) interfaces are in the early phases of standardization.

The PMD bit transmission capability includes bit transfer and bit alignment. It may also include line coding, opto-electronic conversion, modulation and demodulation functions necessary to transfer bits over a particular medium. Because all of these functions depend on the physical medium characteristics, they differ from medium to medium. Hence, this sublayer is referred to as the Physical Medium Dependent or PMD sublayer. It is important to note that the PMD sublayer is only responsible for carrying raw information, in the form of bits. All bits are identical to the PMD sublayer whether voice, data, video, framing, OA&M, overhead, etc.

Because the PMD sublayer provides a logical bit interface to the TC sublayer, the TC sublayer is shielded from the details and characteristics of the physical medium being used. This enables the TC sublayer to be specified independently of the underlying physical medium, allowing it to operate over different media. The TC sublayer performs five functions associated with the clocking of the bit stream onto the physical medium and ATM cells. The lowest function is the generation and recovery of the transmission frame.

Transmission frame adaptation is responsible for adapting the cell flow according to the payload structure of the transmission system in the sending direction. In the reverse direction, it extracts the cell flow from the transmission frame. (In systems deployed before ATM cells, Time Division Multiplexing (TDM) frame format as defined by the SONET/synchronous digital hierarchy (SDH) envelope is used.)

Cell delineation is the mechanism that enables the receiver to recover the cell boundaries. On reception, the TC sublayer must delineate the individual cells in the received bit stream, either from the TDM frame directly or via the HEC in the ATM cell header. To protect the cell delineation mechanism from malicious attack, the information field of a cell is scrambled before transmission. On reception, descrambling is performed.

HEC sequence generation can detect and correct errors on the transmit or receive side. The HEC sequence is generated in hardware and inserted in the appropriate ATM cell header field. At the receiving end, the HEC value is recalculated and compared to the received value. The HEC sequence algorithm is designed so that it can detect and correct single errors and also detect double errors. However, it is not powerful enough to correct double errors. If errors can not be corrected, the cell is discarded.

The *cell rate decoupling* mechanism inserts idle cells in the sending or transmit direction in order to adapt the rate of ATM cells to the payload capacity of the transmission system. This is important when the ATM layer has not been provided with information for transmission, and it allows the ATM layer to operate with a wide range of different physical layer interfaces. In the receive direction, this mechanism suppresses all idle cells. Only assigned and unassigned cells are passed to the ATM layer above. The physical media that have synchronous time slots (DS1/E1, DS3/E3, SONET, SDH, STP, and Fiber Channel-based method interfaces) require this function for decoupling and speed matching. Asynchronous media, such as FDDI PMD, do not. Essentially the transmitter multiplexes multiple VPI/VCI cell streams, queuing them if an ATM slot is not immediately available. If the queue is empty when the time arrives to fill the next synchronous cell time slot, then the TC sublayer inserts an unassigned or idle cell. The receiver extracts unassigned or idle cells and distributes the other assigned cells to the higher layer at the destination.

The application and descriptions of these physical layer functions to the SONET/SDH implementation is discussed in more detail in Part II.

There has been some criticism regarding the proliferation of physical interfaces. Perhaps this stems from the false assumption that there should be only a single standard physical layer interface. However, this would not be a strength but a limitation in the adoption of ATM. Essentially ATM has been adapted to all commonly available physical media with the objective that no re-wiring should be necessary. The large number of physical interfaces is a strength allowing ATM protocols to operate in different environments from the desktop, LANs, campus networks and WANs. The ability of ATM to utilize the existing wiring and transmission hierarchy is important, since replacement of legacy wiring and associated labor costs represent the largest expense.

6 ATM LAYER

The next layer going up the Protocol Reference Model is the ATM Layer. Details of the ATM layer including the cell structure, virtual channel concept, multiplexing and switching were described more fully in Section 2. Section 6 is provided for continuity in describing the levels of the protocol stack.

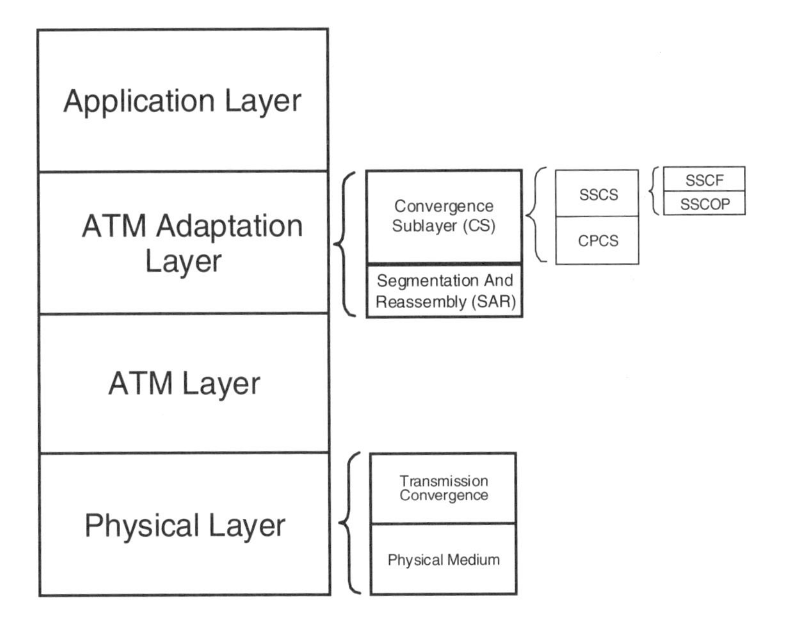

CPCS: Common Part Convergence Sublayer

SSCS: Service Specific Convergence Sublayer

SSCOP: Service Specific Connection Oriented Protocol

SSCF: Service Specific Coordination Function

ATM Protocol Reference
Fig. 6-1

7 ATM ADAPTATION LAYERS (AAL)

The ATM Adaptation Layer (AAL) is the third layer of the simple ATM protocol model. While all information is carried by the same 48-octet ATM cell payload, not all information comes in a 48-octet size. The AAL was developed as a protocol layer for mapping different higher layer information types into the lower ATM layer. It also provides functions needed to "adapt" the services provided by the ATM Layer to the services required by higher layers, generally the service carried/offered to the user. It provides the interface between the user applications and the ATM layer. Figures 7-1 and 7-2 illustrate the mapping of user application information into ATM cells. The AAL may be divided into two sublayers, the Segmentation and Reassembly (SAR) sublayer, and the Convergence Sublayer (CS).

The SAR is responsible for dividing or segmenting the outgoing user information into the proper 48-octet payload of the ATM cell for transport and reassembly of incoming user information into the application/service format on the receive side. The 48-Byte payload is also called SAR Protocol Data Unit (SAR-PDU). This data field consists of a SAR header, a SAR trailer and a SAR Service Data Unit (SAR-SDU). Not every AAL type requires both the header and the trailer.

The CS Sublayer is application/service dependent, and is further divided into *Service Specific Convergence Sublayer* (SSCS) and *Common Part Convergence Sublayer* (CPCS)

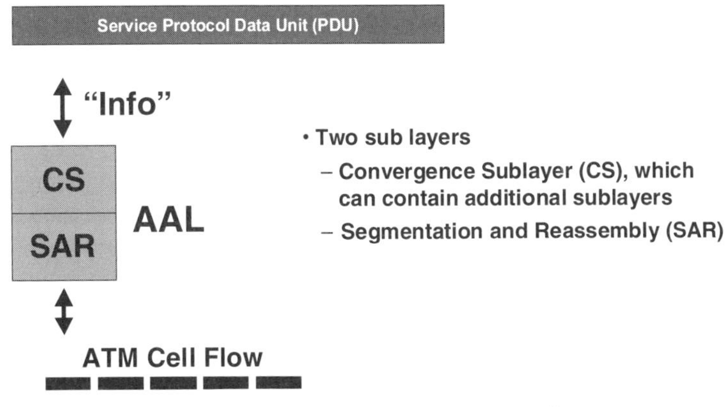

ATM Adaptation Layer (AAL) Functions

Fig. 7-1

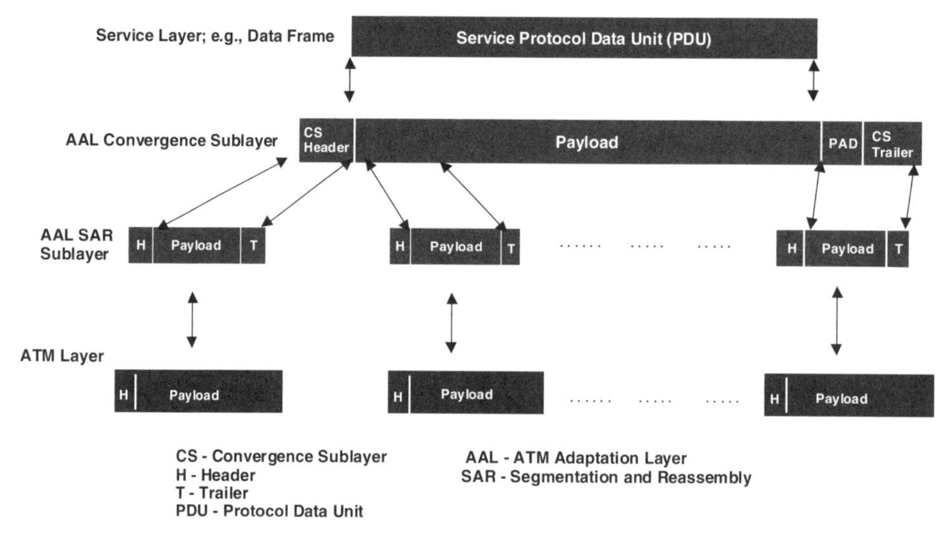

Mapping User Data to ATM Protocols
Fig. 7-2

components. The SSCS sublayer may not be required and can be null. CPCS functions are AAL type dependent and are discussed below. SSCS may be needed for connection-oriented services. The SSCS is further divided into two sublayers, the Service Specific Connection Oriented Protocol (SSCOP) and the Service Specific Coordination Functions (SSCF). See Figure 7-3.

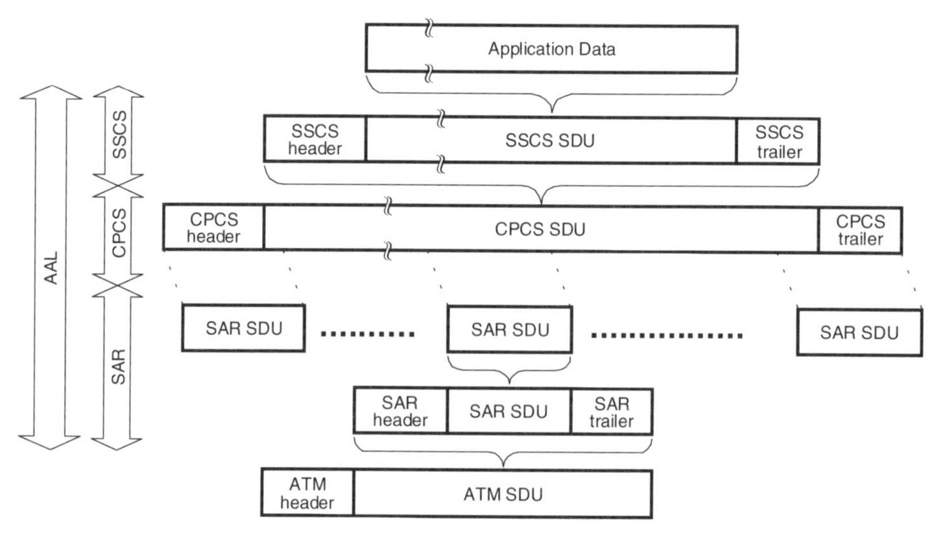

AAL Sublayer Architecture
Fig. 7-3

7.1 CORRELATING AAL TYPES WITH SERVICE CLASSES

Services are broadly characterized by the way information flows: Constant Bit Rate (CBR) or Variable Bit Rate (VBR). AAL adapts the application data to the appropriate connection type, QoS, and in some cases, traffic type, to ATM. With CBR, the user is communicating with a continuous or guaranteed stream of information for an application where end-to-end synchronization or timing is critical. In VBR applications, the source is sending information in a bursty and random fashion. This is typically used in packet data communications. VBR packet communications operate in one of two modes: connection-oriented or connectionless.

VBR packet connections may utilize different traffic types and can be characterized with any or all of the following attributes: peak, average and minimum bandwidths specified along with other QoS parameters if desired for guaranteed level of service. VBR provides transport with variable bandwidth demand. This typically is used for video conferencing applications and transports data that remains after CBR applications are transported.

A committed bit rate guarantees a user that a specific amount of bandwidth will be available regardless of how full the bandwidth pipe is. CBR is the highest guarantee on a shared link. It is often used for voice and video applications. Unspecified Bit Rate (UBR) uses whatever bandwidth is available providing "best effort" services for applications such as e-mail, file transfers and IP/Internet applications. Available Bit Rate (ABR) identifies how much bandwidth is remaining and available for use by applications. ABR was developed for LAN and TCP/IP environments that need more assurance of how much bandwidth is available.

Guaranteed Frame Rate (GFR) is a new traffic classification completed and approved during 1999. It provides some minimum bandwidth throughput. GFR is briefly described in the Traffic Management section of this book.

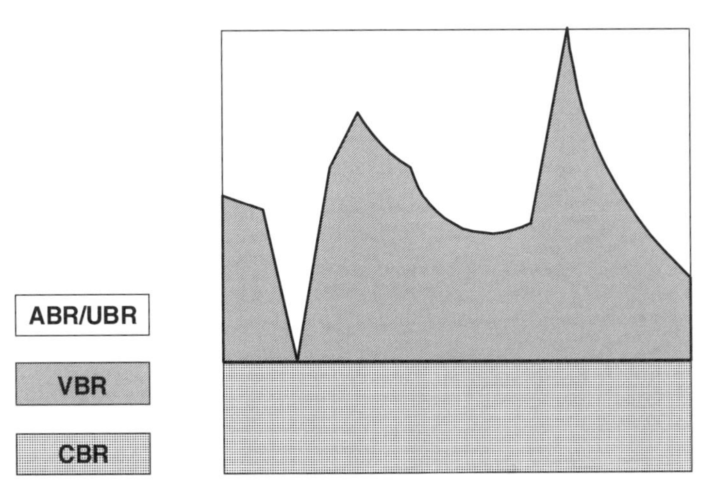

AAL Sublayer Architecture
Fig. 7-4

Figure 7-4 illustrates that ATM manages the bandwidth for the different types of QoS.

Because ATM supports the transport of several different connection and traffic types, and service types, a service classification was developed. The classification categories are based on the timing relationship between source and destination, bit rate, and connection mode. Service classes and the corresponding AAL protocols are listed here:

- *Class A:* constant bit rate service with end-to-end timing. Connection-oriented mode of communication. Applications include voice, circuit emulation (for example, transport of T1/E1 transmission rate of 1.544 Mbps/2.048 Mbps or T3/E3 rates), and video
- *Class B:* variable bit-rate service with end-to-end timing. Connection-oriented communication. Expected applications may be packet video, audio/voice.
- *Class C:* variable bit-rate service with no timing required. Connection-oriented. Example applications are user signaling, Frame Relay, and X.25.
- *Class D:* variable bit-rate service with no timing needed. Connectionless mode. Applications include IP, SMDS.
- *Class X:* null AAL or not specified and called "Native ATM." User provides AAL, can be any existing or proprietary AAL.
- *Class Y:* used to identify VBR ABR connection.

Parameter	Class A	Class B	Class C	Class D	Class X "Native ATM"	Class Y
Timing Relationship between source and destination	Required		Not required			
Bit Rate	Constant	Variable				
Connection Mode	Connection oriented			Connectionless	Connection Oriented	
Supporting AAL Type	AAL 1 Or **AAL 5***	AAL 1, AAL 2, or AAL 5 (+)	AAL 3/4, **AAL 5**	AAL 3/4 Or **AAL 5**	Any type — User defined	AAL 3/4 **AAL 5**
Associated QoS	CBR	VBR	VBR	VBR	UBR	ABR

*Note 1: AAL 5 has recently been specified for voice support as defined in the ATM Forum Voice and Telephony Over ATM (VTOA) version 1.0 specification with AAL 1 as optional and in the Audio/visual Multimedia Service (AMS) version 1.0 specifications. ITU-T requires AAL 1 for voice while AAL 5 is optional.

Note 2: **AAL 5 highlighted in bold** identifies the preferred AAL type based on ATM Forum specifications. The objective is to ease future migration and upgrades to all VBR/packet network infrastructure.

(+) Note 3: Industry and standards activities are currently underway. It is premature to determine what will emerge as the preferred AAL for Class B service connections.

Services Classes Transported Over ATM

Table 7-1

Four AAL types were originally created, AAL 1, AAL 2, AAL 3, and AAL 4. AAL 3 and AAL 4 are merged into AAL 3/4 during the development and approval of the original 13 ITU-T ATM Recommendations approved in November, 1990. AAL 5 was introduced after the approval of the recommendations and approved approximately two years later. The service classes are defined by timing relationship between source and destination, bit rate, and connection mode. Table 7-1 summarizes the service classes and which AAL types support them.

7.2 AAL TYPES

The choice of AAL depends on the service. The AAL is implemented in the end user's equipment and may also be terminated in the network depending on the service offered by the network. Currently four AAL types are specified in standards, along with the null AAL or "Native ATM," Figure 7-5.

7.2.1 AAL 1

AAL Type 1 is used for the primary applications: carrying synchronous CBR

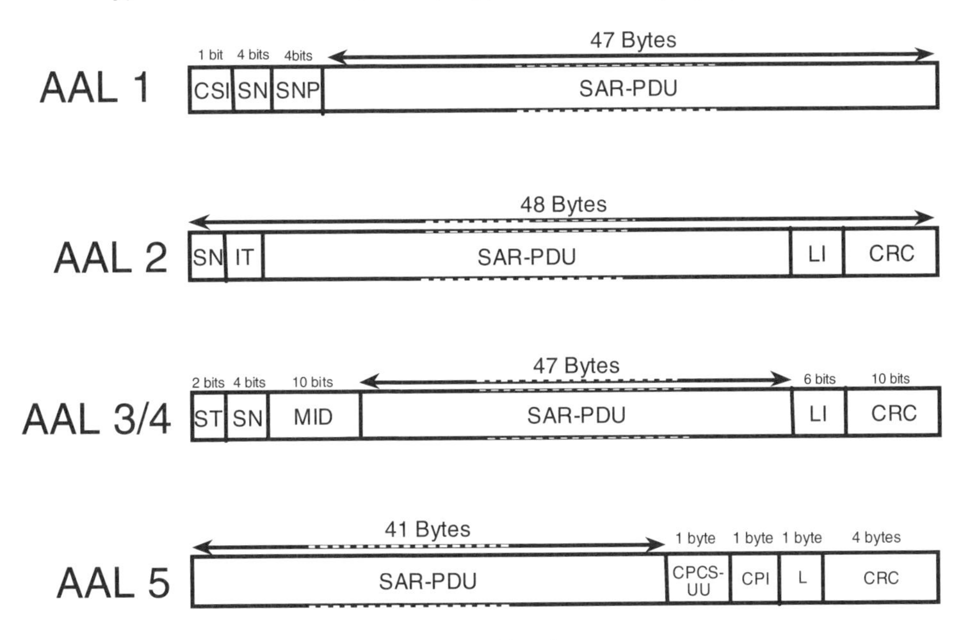

CPCS-UU: Common Part Convergence Sublayer- User-User
CPI: Common Part Indicator
CSI: Convergence Sublayer Indicator
CRC: Cyclic Redundancy Check
IT: Information Type
LI: Length Indicator
MID: Message Identifier
SAR-PDU: Segmentation And Reassembly-Protocol Data Unit
SN: Sequence Number
SNP: Sequence Number Protection

AAL Types
Fig. 7-5

applications, such as voice; transparent transport of T1/E1 and T3/E3 through the ATM asynchronous network using the service capability known as Circuit Emulation Service (CES); and real time dependent video and television. Providing Class A service means that, at the transmitting ATM endpoint, the AAL 1 service must accept data from the CBR application continuously at a constant rate or constant stream of traffic. At the receiving endpoint, the AAL 1 service must also present the same bit stream at a constant rate to the AAL 1 user. Since there is jitter in the ATM network that is introduced by buffering and multiplexing at each ATM switch along the path, the AAL 1 receiver must perform jitter removal (for example by using a smoothing buffer) to maintain a constant rate.

The AAL 1 has a fixed size PDU that can be carried in exactly one cell. This means that the SAR sublayer does not need to perform a segmentation or reassembly function since the AAL 1 payload fits into one cell. The SAR-PDU consists of a 1-Byte header and 47 octets of SAR-SDU. The header consists of a 1-bit CS Indicator (CSI) field, a 3-bit Sequence Number (SN) field and a 4-bit Sequence Number Protection (SNP) field. The CSI carries the CS indication. The sequence number is used to detect loss and misinserted cells. The sequence number protection field is used to detect bit errors. The AAL 1 CS function includes handling Cell Delay Variation (CDV), handling cell payload assembly delay and recovering the source clock at the receiver. At the receiving end, the AAL is terminated and the CBR stream is reconstructed. See Figure 7-6. Then, the source clock is recovered, the jitter accumulation through the transit network is eliminated, cell sequence is verified, and error checks are performed.

AAL 1 must maintain the same data rate to the higher layer on the receiving endpoint to avoid buffer overflow or underflow over a period of time. Rate mismatch, network jitter or differences in clock recovery between the sender and receiver contribute to buffer overflow and underflow. When the ATM cells are arriving so fast that the buffer may overflow, the receiver side AAL 1 layer is responsible for dropping the ATM cells with an indication to the AAL 1 user.

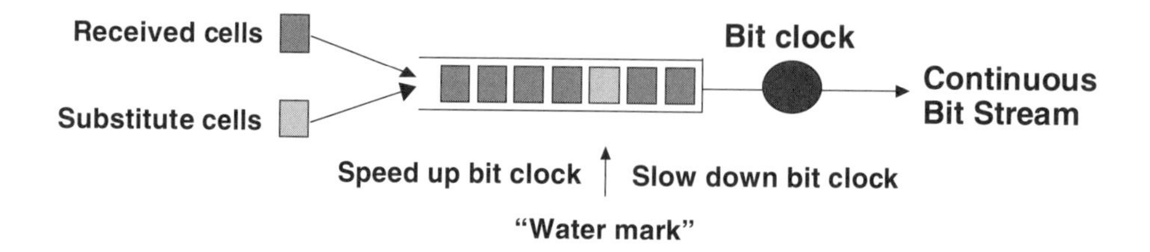

- Bit stream rate is independent of ATM network
- Cell delay variation critical to buffer sizing and bit clock jitter
- Parameters standardized for *n* x 64 Kbps up to 1.39264 Mbps

AAL 1 Adaptive Clock Recovery Method
Fig. 7-6

Conversely, if there is a buffer underflow, AAL 1 is responsible for substituting a cell, as seen in Figure 7-6, to maintain a constant rate to the AAL 1 user.

7.2.2 AAL 2

AAL Type 2 is used to handle VBR traffic where a strong timing relationship needs to exist between the source and the destination, but the bit rate and PDU size may vary, such as VBR packet voice or video. AAL 2 was approved as a standard by ITU-T in September, 1997. The SAR-PDU consists of a variable header, trailer and SAR-SDU. The SN is used to detect loss and misinserted cells. The Information Type (IT) is used to indicate the beginning of a message (BOM), the continuation of a message (COM) and the end of a message (EOM). The Length Indicator (LI) indicates the length of useful octets in a partially filled cell. The Cyclic Redundancy Check (CRC) is used to correct bit errors in the SAR-PDU.

The AAL2 CS functions include clock recovery, handling lost or misinserted cells and Forward Error Correction (FEC). The application and higher layer protocol functions to use AAL 2 have not yet been fully defined. To date, most of the ITU-T standards and ATM specification work is focused on VBR-packet voice driven by the PBX and enterprise network industry segments for lower cost long distance connections. The objective is to allow multiple voice conversations to be carried in one ATM cell. With low bit rate voice coding techniques, only 8 or 16 octets of information may be required to carry speech. The mapping of information (voice samples) from more than one user into a single ATM cell is referred to as a composite cell.

With composite cells, all information must go between the same two user destination points. At this point in time, most ATM switches do not "look into" an ATM cell to determine where to switch the cell. To do so would slow down switching performance.

The use of AAL2 is also of interest to the wireless ATM industry segment. The use of AAL2 may enable a single ATM cell to carry a voice sample along with either fax or IP/Internet data to a multimedia wireless ATM user. However, to enable this to occur requires a significant amount of additional standardization development. This involves user and application addressing/numbering, signaling and control protocol extensions, and multimedia protocols defining higher layer functions to handle the mix of media information.

7.2.3 AAL 3/4

AAL 3/4 was developed for transmitting VBR information where the network cell delay and its variation have no deleterious real-time effects and cell loss is less critical because of the recuperative effects of the higher layer application protocols applied by the user.

AAL 3/4 is used in connectionless Switched Multimegabit Data Service (SMDS) applications [18]. SMDS provides reliable VBR packet data service, and some multimedia experimental applications involving voice and MPEG coded video.

The SAR-PDU consists of a 2-Byte header, a 2-Byte trailer and a 44 Byte SAR-SDU, Figure 7-7. The header consists of a 2-bit Segment Type (ST) field,

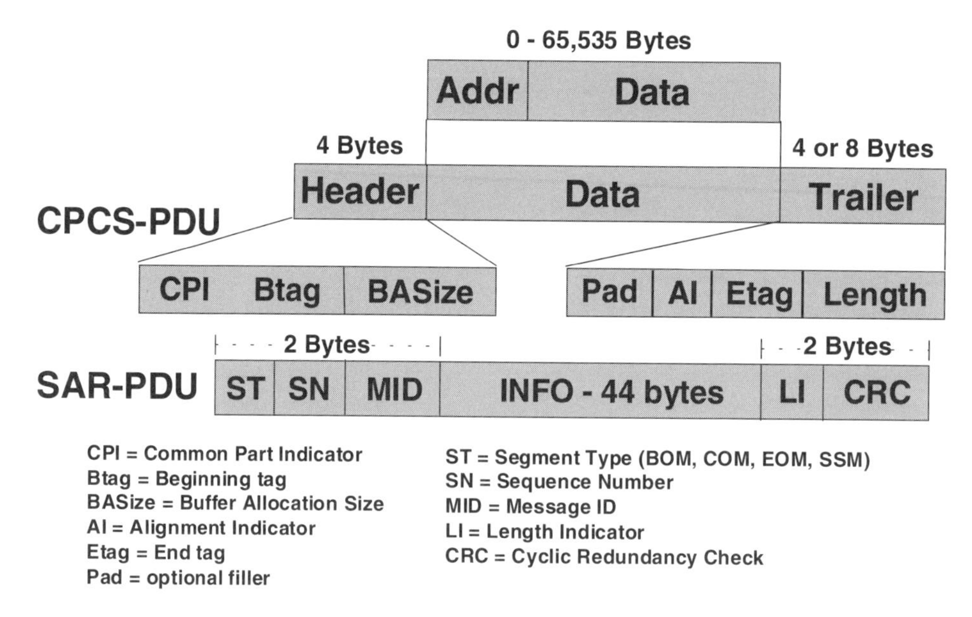

AAL 3/4 Format
Fig. 7-7

a 4-bit Sequence Number (SN) field and a 10-bit Multiplexing Identifier (MID) field. The trailer consists of a 6-bit Length Indicator (LI) field and a 10-bit CRC field. The ST field indicates Beginning Of Message (BOM), Continue Of Message (COM), End Of Message (EOM) and Single Segment Message (SSM). The SN is used to number the SAR-PDU from 0 to 15. The MID is used for multiplexing AAL connections into a single ATM connection for connection-oriented data communication, and is well suited for multimedia applications. The LI indicates the length of useful octets in a partially filled cell. The 10-bit CRC is on a per cell basis. The CRC is used together with the SN to provide a very powerful bit error protection and to correct bit errors over the entire SAR-PDU. This is particularly useful for applications that can not have retransmission. However, for those applications that already provide retransmission by higher layers of the protocol, such as TCP on top of the IP protocol, this can be overkill for a low bit error rate environment as experienced with fiber transmission facilities. (This was part of the motivation that led to the development of AAL5 described in the next section.)

The CS functions include CPCS Service Data Unit (SDU) preservation, error detection and handling, buffer size allocation and partially transmitted CPCS-PDU abortion. The CPCS-PDU format for AAL 3/4 is shown in Figure 7-8. The CPCS-PDU payload length can range from 1 to 65,535 octets. The CPCS header consists of a 1-Byte Common Part Indicator (CPI) field, a 1-Byte Beginning tag (Btag) field and a 2-Byte Buffer Allocation Size (BASize) field. The CPCS trailer consists of a 1-Byte Alignment (AL) field, a 1-Byte End tag (Etag) field and a 2-Byte Length field. The padding field ranges from 0 to 3 octets and is between the CPCS-PDU payload and the CPCS trailer. It is used

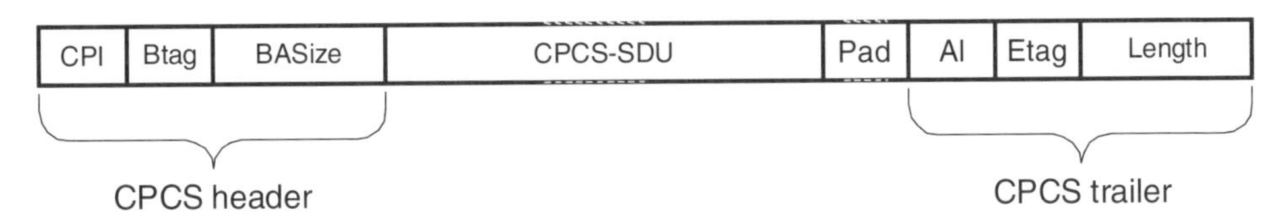

AAL 3/4 CPCS-PDU Format
Fig. 7-8

to stuff the last CPCS-SDU making up the CPCS-PDU to be the correct length which is a multiple of 4-octets.

The CPI indicates the counting units for the values of BASize and length fields. The Btag and the Etag fields are assigned the same value at the transmitting end, and the values are matched at the receiving end for each CPSC-PDU. BASize indicates the maximum buffer required to receive the CPCS-PDU. AI is used to stuff the CPCS trailer into the correct length which is a multiple of 4 octets, similar to the padding field. The values for these two fields are zeros. The length field indicates the length of CPCS-PDU user information.

7.2.4 AAL 5

AAL 5 was originally developed to carry VBR traffic between computers, in either connection-oriented or connectionless environment, with a simple encapsulation of the packets generated by the source. Applications using AAL 5 include multiprotocol (including IP) operation over ATM, LAN Emulation (LANE) data, database query, and Frame Relay (FR) service. AAL 5 is also used to support user signaling and network signaling protocols. In addition, AAL 5 supports CBR connections. Support for connectionless or connection-oriented services and CBR applications is provided by the Service Specific Convergence Sublayer (SSCS) level. For example, the ATM Forum developed specifications for Audiovisual Multimedia Service (AMS) [19] supporting MPEG 2 video for video-on-demand distribution services and future video conferencing, and the Voice and Telephony Over ATM (VTOA) to Desktop [20] using AAL 5 in CBR[8] connections. The specifications include the definition of the higher layer protocol functions necessary to use AAL5 with CBR mode connections.

The objective for AAL5 development was to design a simple AAL type that is efficient in terms of processing overhead and cell transmission overhead, especially for PC end stations. The original proposal[9] was called the Simple and Efficient Adaptation Layer (SEAL). Although AAL3/4 had been defined to support the same classes of service, the computer industry segment concluded that AAL3/4 incurs too much processing and transmission overhead for PCs. Many of the AAL3/4 functions such as multiplexing, per cell CRC, sequence numbering and buffer allocation are eliminated for AAL5.

AAL5 has another unique characteristic. Instead of using a per cell indication of its position in the AAL5 PDU (such as a sequence number or indicating

[8] Use of AAL5 in CBR mode was done to ease VTOA migration and interoperability with the VTOA VBR/packet target objective. The VBR/packet version of voice is under development.

[9] T. Lyon, "Simple and Efficient Adaptation Layer (SEAL)," ANSI contribution T1S1.5/91-292, August 1991.

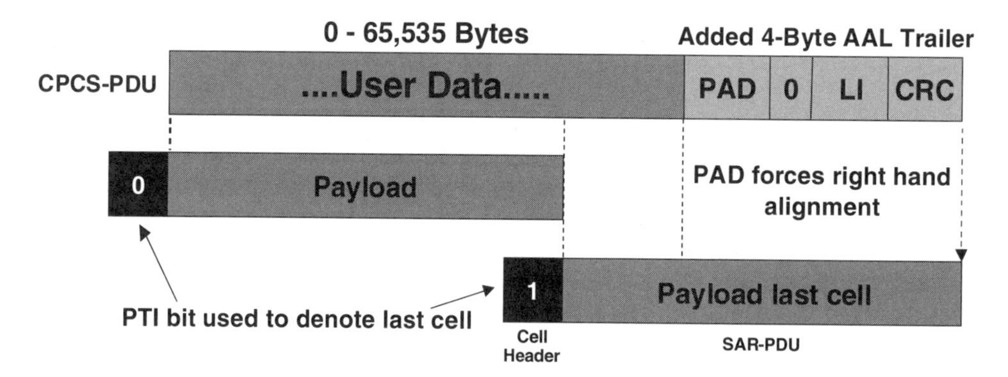

- Non-zero value in AAL - indicate in ATM cell header PTI field identifies the last cell of the sequence of cells making up the PDU
- Receiver simply places cell payloads into memory until "last cell" indication occurs
- It then checks the 4-Byte CRC, compares LI to received length

AAL5 Format
Fig. 7-9

whether it is the beginning, middle or end of a packet) only the end of an AAL5 PDU or the last cell is identified. See Figure 7-9. AAL5 takes advantage of the PTI (Payload Type Indicator) field in the ATM cell header as a flag to indicate that this ATM cell is the last cell of an AAL5 PDU. In general, the AAL5 SAR-PDU has no header and trailer overhead bits on a regular basis. The last ATM cell of the AAL5 PDU consists of a 41-Byte SAR-SDU and a 7-Byte CPCS trailer. The AAL5 SAR functions include SAR-SDU preservation, congestion information handling and loss priority information handling.

The use of an indicator flag in the ATM cell header provides certain other advantages from an overall network efficiency perspective. Advanced congestion features can be implemented. During congestion periods, for example, it is more effective to selectively drop cells that belong to a particular AAL5 PDU or limited number of cells rather than randomly dropping ATM cells across many different PDUs. In data communications applications, a cell loss means that the entire PDU is useless and a retransmission of the entire PDU is needed. By minimizing the number of PDUs with cell losses means that the number of PDU retransmissions across the congested ATM network can be minimized. Further, ATM switches can check the ATM cell headers during facility congestion and can decide that all the cells belonging to an AAL5 PDU that are buffered and awaiting transmission can be dropped or flushed up to the last cell. This helps avoid transmissions that may be contributing to the congestion conditions.

The CS functions include CPCS Service Data Unit (SDU) preservation, error detection and handling and partially transmitted CPCS-PDU abortion. It does not support buffer size allocation function. The CPCS-PDU payload length also can range from 1 to 65,535 octets. The CPCS-PDU trailer consists of a 1-Byte

CPCS User-User information (CPCS-UU) field, a 1-Byte Common Part Indicator (CPI) field, a 1-Byte length field and a 4-Byte CRC field.

The CPCS-UU transfers CPCS user-to-user information transparently. The CPI interprets subsequent fields for the CPCS trailer. The length field indicates the length of CPCS-SDU. The CRC-32 detects errors in the CPCS-PDU. The CRC-32 is the same one used in IEEE 802.3, IEEE 802.5 LAN standards, FDDI and Fiber Channel.

7.2.5 Null AAL or Class X

Null AAL in support of Class X service is also known as "Native ATM" or raw ATM layer service. No higher layer AAL support services are provided by the ATM network in these applications. This case is referred to a null AAL which provides basic functions of ATM switching and transport and is called Cell Relay Service (CRS). This provides flexibility, particularly for users who want to provide their own proprietary AAL for special applications or experimentation. In these cases, the AAL involves only the CPE at both ends of the connections and is transparent to the network. Class X or null AAL is a connection-oriented service. The traffic characteristics can be either CBR or VBR and the timing requirements between the sender and destination are all user defined and transparent to the ATM network. Class X users still specify the bandwidth and QoS requirements in the signaling connection setup messages.

8 CONTROL PLANE ADDRESSING

The Control Plane provides virtual connection control related functions. The most important of these functions is the Switched Virtual Connection (SVC) capability. To enable SVCs, the control plane provides the critical addressing and routing mechanism as well as signaling.

8.1 ADDRESSING

Addressing is critical in identifying and controlling switched connections. Addressing occurs at two levels; at the ATM VPI/VCI level and the logical network level. Because the VPI/VCI address is unique only to a transmission path, there is a need to have a higher level address that is unique across each network (see Figure 8-1).

8.1.1 ATM Layer VPI/VCI Addressing

The signaling protocol assigns VPI/VCI values to ATM addresses and physical ATM UNI and/or NNI ports based on a number of factors. These factors include type of SVC connection requested, resources available and whether the subscriber is entitled to the requested service. These decisions are made by the switch call processing and routing software. A physical ATM UNI port must have at least one unique ATM address and may have more addresses as defined by the maximum number of connections defined by the VPI/VCI

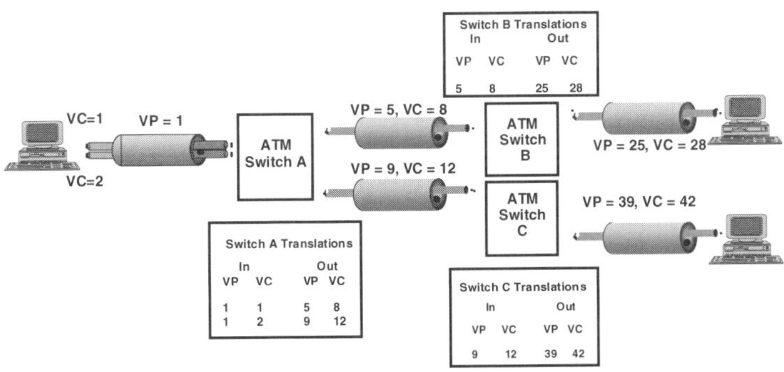

VPI/VCI Addressing and Switching

Fig. 8-1

field. Because a Virtual Channel Connection (VCC) and Virtual Path Connection (VPC) are uni-directional, meaning the traffic parameters for the forward direction of the connection and the return direction are independent, the signaling protocol must specify traffic parameters for each direction as well as the VPI/VCI address. The switch call processing system is responsible for checking subscription information and resource management while the routing software is responsible for finding a path.

8.1.2 ATM SVC Network Level Addressing

The objectives of addressing are to enable construction of a scalable address structure, ensure that the addressing scheme is easy to administer, provide the ability to uniquely identify an ATM endpoint, and to accommodate public/private interworking using the existing technology where appropriate. There are two types of ATM control plane addressing plans to support establishing end-to-end Switched Virtual Connections (SVC) through networks. The ITU-T Recommendation E.164 has been the basis of telephony addressing/numbering plans around the world for identifying narrowband and broadband users. The E.164 format is useful for organizations that may wish to use the existing, largely geographic-based public telephony numbering format. The North American Numbering Plan (NANP) is based on E.164 with some modifications specific to North American dialing services. ATM supports the use of NANP based numbers referred to as "native" E.164 and NSAP E.164[10] addresses. The public (ATM) address is communicated in the signaling Called Party Number information element. However, industry participants believed the existing public network-based addressing scheme was not sufficient to address private or enterprise network needs and use of IP/Internet-based addressing within corporate networks. The ATM Forum has developed an alternative addressing/numbering plan called *ATM End System Addressing* (AESA) that is a superset of the current public addressing/numbering plan introducing new private networking addressing capabilities.

AESA includes support of E.164 address format to ensure compatibility with today's public network, covering services ranging from Plain Old Telephony Service (POTS), to faxes, to Narrowband ISDN data and voice services. Individual ATM interfaces can have more than one E.164 associated address. The address translation procedures associated with telephony numbering apply to E.164, so that we might expect these ATM addresses to show the same sort of advanced address services common in telephony, such as 800 number translation and billing, 900 number services, credit card calling, and others.

The intersection of private ATM addresses and public networks is messy. Resolving the technical interworking of address formats is the simplest part of the challenge. More complicated is the matter of reconciling the address service models which derive from the separate roots of telephony, private data networks, and digital video services, and regulatory issues associated with the allocation of addresses.

---

[10] Native E.164 and NSAP E 164 addressing refers to the two ways that the E.164 "telephone number" is coded and carried in signaling messages.

20 Bytes
AFI (1B)
ESI (6B)
SEL (1B)

AFI = Authority and Format Identifier
ESI = End System Identifier
SEL = Selector

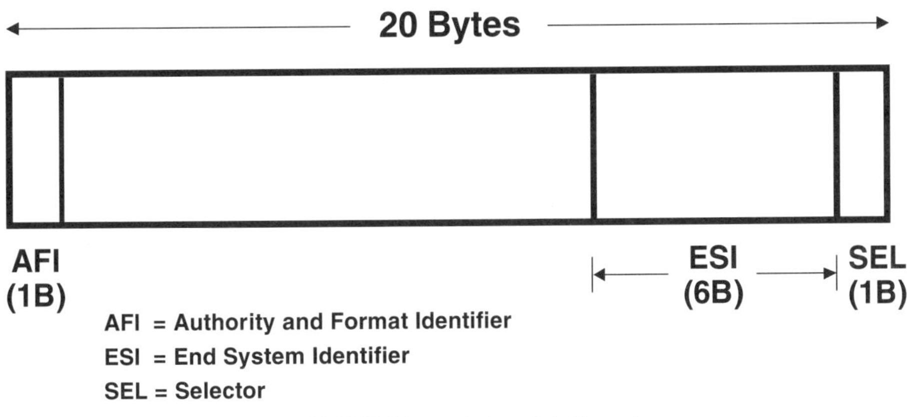

ATM SVC Network Level Addressing
Fig. 8-2

The Private ATM Address format is actually a family of three different formats. All have a common structure (Figure 8-2) with a 20-Byte length, including a leading Byte to identify the detailed format, a 6-Byte End System Identifier, and a 1-Byte Selector field. The 6-Byte ESI field is generally considered to be an IEEE MAC address, thus a hardware tag is generated which is guaranteed to be unique within a routing domain.

The SEL *(Selector)* field usually is not used in routing, but can be used by the end system. One example frequently cited is that the SEL field might be used to connect to an end system application or process.

The AFI field is the *Authority and Format Identifier*, a tag used to identify which authority is managing the following address space and implicitly defines the format of the remainder of the address. Three such AFIs are currently defined; and they yield the three private ATM address formats.

The first format of Figure 8-3, when AFI=39, is Binary Coded Decimal. This

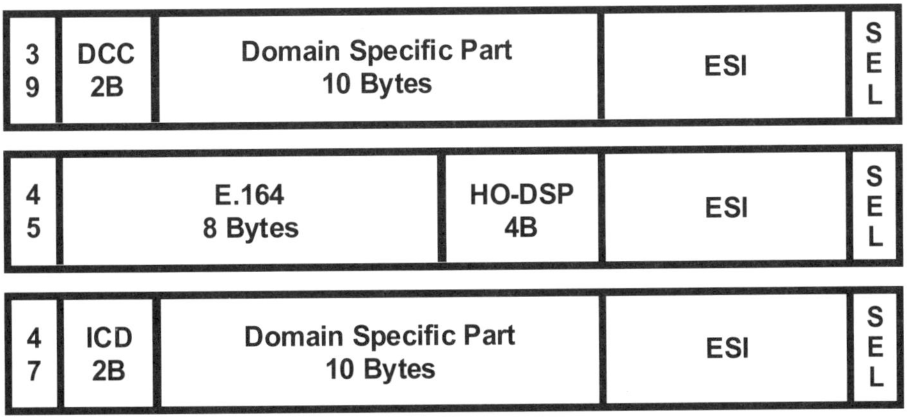

39
DCC 2B
Domain Specific Part 10 Bytes
ESI
SEL
45
E.164 8 Bytes
HO-DSP 4B
ESI
SEL
47
ICD 2B
Domain Specific Part 10 Bytes
ESI
SEL

Private ATM Addresses: Specific Format
Fig. 8-3

means that the address is formatted according to the *Data Country Code* format, with a 2-octet *Country Code* indicating which national authority owns the remaining 10 Bytes of the address. These country codes are spelled out in ISO 3166.

The second format, indicated when AFI = 45, is an E.164 AESA address, encoded in 8 Bytes with a 4-Byte free field for additional routing capability and the 6-Byte hardware address.

The third format, indicated when AFI = 47, uses the *International Code Designator* to identify an international organization. The British Standards Institute makes these identifiers available to international organizations. 10-Bytes are available for hierarchy which is assigned below the organization code.

The choice of an actual addressing scheme for any particular enterprise is not a simple one. Large multinational companies may choose to use the ICD format for their private networks.

How are ATM interface numbers determined? The assignment and administration of E.164 AESA is done separately, with the E.164 address component determined by the public network service provider and the remaining fields handled by the private network operator. The public E.164 address is administered by the numbering authority for that world zone. The non-E.164 number is assigned in whichever fashion the ICD- or DCC- identified organization chooses.

Address registration: Addressing registration refers to a procedure that would allow a user to move their terminal equipment from one location, plug it into the appropriate connector, and upon initialization, commence communication without requiring a manual service order process. For purposes of address registration, an ATM address is divided into two parts: a "user part" and a "network prefix." The user part identifies the ESI and SEL fields of a private AESA; all other fields constitute the network prefix. For a native E.164 address, the network prefix is the E.164 address and the user part is null. The user part is supplied by the user-side of the UNI, while the network prefix is supplied by the network-side. Each UNI can support more than one network prefix. However, the SEL is irrelevant for address registration; multiple user parts that differ in only the SEL are treated as duplicate addresses.

The address registration procedure is initiated by the switch sending the network prefix(es) to the user side of a UNI. The user side completes the registration process by appending its user part to the network prefix and sending the entire address back to the switch. The user-side can repeat this process for each user part that needs to be associated with a single network prefix.

Benefits from the address registration procedure include:

- Identifying addresses at initialization of a UNI, and de-registering addresses when the link goes down
- Reflecting changes in the addresses associated with a UNI while the link is active.
- Managing the addresses that can be used on the UNI (e.g., blocking the registration of an address that has already been registered on a different UNI).

Address registration is accomplished via the Integrated Local Management

Interface (ILMI). Two new Managed Information Base (MIB) object tables have been defined as part of the ILMI network management specifications. These are:

- Network Prefix table (to be implemented on the user side)
- Address table for complete ATM addresses (to be implemented on the network side)

Group Address: An ATM "group" is a collection of ATM end-systems that may have one or more members. Currently, the only method defined for creating groups and adding (or subtracting) members is to use the ILMI address registration procedures. A single end-system can be a member of different groups simultaneously. An ATM "group address" is assigned to identify the group, in contrast to the "individual address" which has a one-to-one correspondence with a particular UNI. The type of AESA is determined by its AFI: 10-99 (hex) identify individual addresses, while A0-F9 (hex) identify matching group addresses.

If a group address is assigned to a specific service, it then becomes a "well known" group address. This "well-known" group address points to one or more individual addresses of destinations that offer the service. A client can gain access to a service without having to know the actual destination address of the server. Having to keep track of fewer addresses may make development and usage of ATM applications easier. The individual address (if known by the user) can still be used to direct a connection to a particular remote destination. A well-known group address can also be viewed as a default address for the service.

The ATM Forum currently has defined well-known group addresses for LAN Emulation and the ATM Name System (ANS). These addresses use the ICD AESA format, where the prefix is "C50079" (for reference, 0079 represents the ICD assigned to the ATM Forum; the associated individual address prefix is "470079"). The general policy in the ATM Forum is to identify a well-known group address in an implementation agreement which is published as a specification.

Addressing-related supplementary services such as calling number identification, call forwarding, sub-addressing and closed user groups are also supported by ATM broadband signaling. Discussions of addressing-based services go beyond the scope of this book.

8.1.3 ATM Name Service (ANS)

The ATM Name Service (ANS) will translate a name to an ATM address, or return a name if supplied with an ATM address. It works with both native E.164 addresses and AESAs. ANS is based on the IETF's Domain Name Service (DNS), enabling existing DNS resource records to be modified to support ATM. This leveraging of DNS is a big advantage for ANS because DNS is widely implemented, and users already are knowledgeable about its domain names. ANS will be eligible for the extensions to DNS that are being considered by the IETF, including security and dynamic updating. It will also be possible to add additional information beyond names and addresses to the resource records.

The way ANS works is that the client establishes an SVC connection to an ANS server and exchanges a set of queries/responses which are carried over the VC using AAL type 5 and the Undefined Bit Rate service. Communication between servers is based on TCP/IP using recursive processing. The address for the ANS server can be obtained either through the use of ILMI, or the ANS well-known group address can be used in anycast mode. Once the VC is established, the client may submit multiple queries to the server. The VC can be released by either the client or the server after a period of inactivity to free up resources for other queries.

Additional details can be found in the following two specifications: ***ATM Forum Addressing: Reference Guide*** and the ***ATM Forum Addressing: User Guide***. The ATM Forum Addressing: Reference Guide describes the new capabilities in addressing ATM networks to locate a user. It also defines AESA formats developed by the ATM Forum, provides an overview and discusses implications associated with AESA alternatives, and presents several case studies on how to apply AESAs. The User Guide provides information on where to obtain ASEA numbers, how they are used in private networks, and how they are used to attach to service providers/operators. Analogies between ATM Forum addressing and IP addressing are also included.

9 CONTROL PLANE SIGNALING

Signaling and control protocols are becoming increasingly important aspects of new digital services for voice, data, video, and multimedia. Signaling represents the act of transferring, in real time, service-related information between the user and the network, and between network entities to establish end-to-end communications. Signaling is important because unless users are able to convey their needs to the network, they will not be able to receive the required services on demand.

Signaling protocols dynamically establish, maintain and clear a connection to a user. Signaling allocates network resources to a connection and enables the network, which statistically aggregates traffic from many customers, to efficiently use its resources. Signaling procedures are defined in terms of call control messages and Information Elements (IEs) that are used to characterize the connection and describe the desired functions to ensure interoperability between end points and between networks. ATM signaling specifies the details of the connection such as the forward and backward bandwidth, the QoS parameters, the type of ATM Adaptation Layer and connection (e.g. CBR, UBR, ABR, and other VBR/packet connections), and interworking, subaddressing, point-to-point, point-to-multipoint connectivity, and other user-to-user information. The signaling messages and IEs must unambiguously and efficiently describe what is needed.

ATM signaling is carried by the same transport systems as the services that utilize the ATM transport capability. However, the signaling and user service flows are logically separated from each other by using different virtual connections with different QoS requirements.

For some time there has been a philosophical debate between "dumb" networks and "smart" networks. This debate is a result of the connectionless-oriented Internet versus connection-oriented telephony networks. In the case of the Internet, most of the intelligence, service/application processing and features must be implemented at the edge and/or in the user terminal equipment. The Internet in this case simply provides "best effort" service by forwarding user information or packets of data to the destination point. If reliable transport is needed by an application, higher layer protocols are layered on top of IP. The Internet network itself is dumb, providing little added value service beyond basic transport.

The telecommunications infrastructure model offers added value services to users by providing service capabilities and functions from the network. These services frequently are number- or address-based and continue to become more sophisticated as new signaling capabilities are put into place. Examples include 800/888 number services, credit card calling, caller identification, time of day routing, support of mobile users, and many supplementary/added value services such voice messaging or call rerouting to another number. These services are connection-oriented.

The tradeoff is between dumb, low function, bit pipe transport versus intelligent network transport. Service providers/operators need to consider the revenue opportunities these fundamentally different architectural approaches offer. Ultimately, explosive growth of data services means that the continued investment in separate voice and data networks is no longer a viable business strategy. Economic principles are driving the convergence of separate voice, video and data networks. For example, the Internet, which has only one level of service - "best effort"- has undertaken standards development efforts to introduce signaling and control protocols so that better levels of service may be provided in the future. The IP based mechanisms are somewhat different than those currently used in narrowband and broadband connection-oriented networks due to connectionless nature. These emerging IP mechanisms are premature to describe because they are still being debated and developed. However, the interworking of these new IP signaling and control protocols will have to be addressed in the future. A bigger fundamental question to be addressed is on

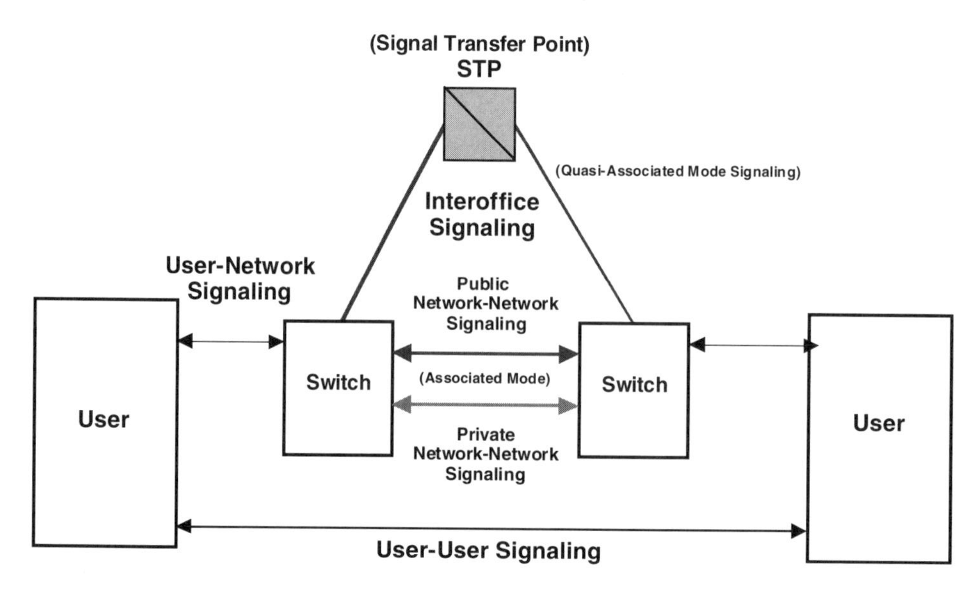

Realms of Signaling
Fig. 9-1

what the next generation network signaling and control protocols will be based: connection-oriented or connectionless protocols, or a convergence of the two.

There are several signaling domains: user-to-network, network-to-network, and user-to-user (see Figure 9-1). Whether public or private, a network would be of limited use unless the user is able to communicate to the network their needs for service. In addition to this user-to-network signaling, there is a need for signaling between various components of the network itself. With ATM, there are two types of network-to-network signaling protocols that have been developed: private network and public network signaling. Differences between these two protocols are discussed later in this section. Finally, once a connection exists to the destination, there may be a need for signaling between the end points referred to as user-to-user, for application specific level functions between terminal devices. Signaling, therefore, refers to the mechanism to establish a connection, to monitor and supervise its status, and to terminate the connection. Also, signals are messages generated by the user or some internal network processor pertaining to call/resource management. Signaling equipment performs the functions of alerting, addressing, supervising, and providing status.

9.1 CONNECTION SETUP

ATM supports a multiplicity of service types and connection types with diverse characteristics. To establish a connection, one of two methods can be used. ATM connections can be set up permanently through a manual service order provisioning process, referred to as *Permanent Virtual Connections (PVCs)*. PVCs, are simple to establish and manage. Resources are dedicated on a continuous basis: 7 days a week, 24 hours a day, whether used or not. This is

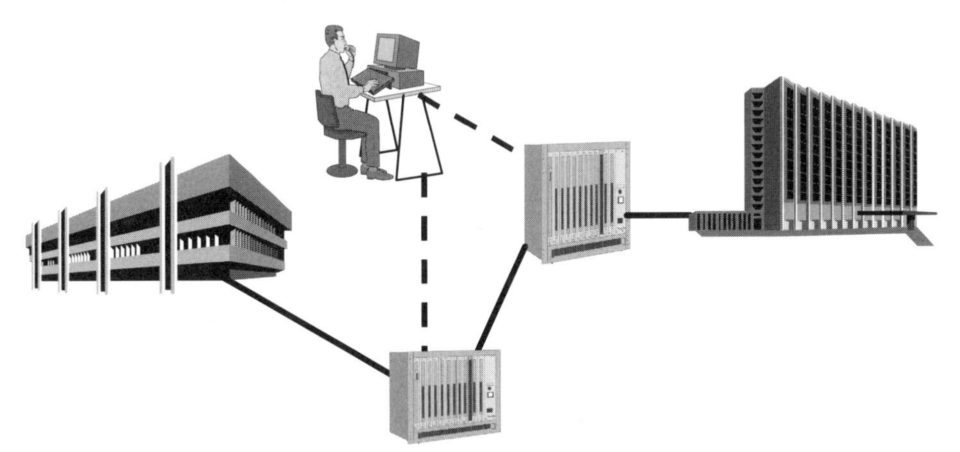

- Established via network management system which sets routing tables in switches
- Assumes long holding time, tolerates long setup time

Permanent Virtual Connections (PVC)
Fig. 9-2

illustrated in Figure 9-2. Part 3, Section 20.1 discusses PVC services and the process involved in a developing Service Level Agreement (SLA) with a customer.

Alternatively, connections can be set up dynamically or on-demand when a connection is needed as shown in Figure 9-3. These connections are referred to as *Switched Virtual Connections (SVCs)*. In these cases, users are billed only for the resources used during the connection.

The mechanism used to communicate between the user and the network as well as establish the type of connection needed and determine the characteristics of that connection is called signaling. The ATM Layer predefines a VC for the use of signaling messages. The signaling from the user into the network is referred to as access signaling and is carried across the UNI.

The signaling between network elements within a network e.g., switches, signaling transfer points, databases, etc., and between different carriers/service providers is referred to as network signaling. Network signaling is carried over a Network Network Interface (NNI) of which there are at least two types. They are discussed later.

To establish a connection, the user communicates to the network via signaling connection **Setup** messages. These messages contain the characteristics of the connection desired and the user destination address. The characteristics are then communicated from the originating network switch through the network to the destination using Setup signaling messages. This allows network resources to be checked for availability. If the resources are available, the switch sends a **Call Proceeding** message back to the network element from which it received the Setup message. The resources are allocated to handle the requested

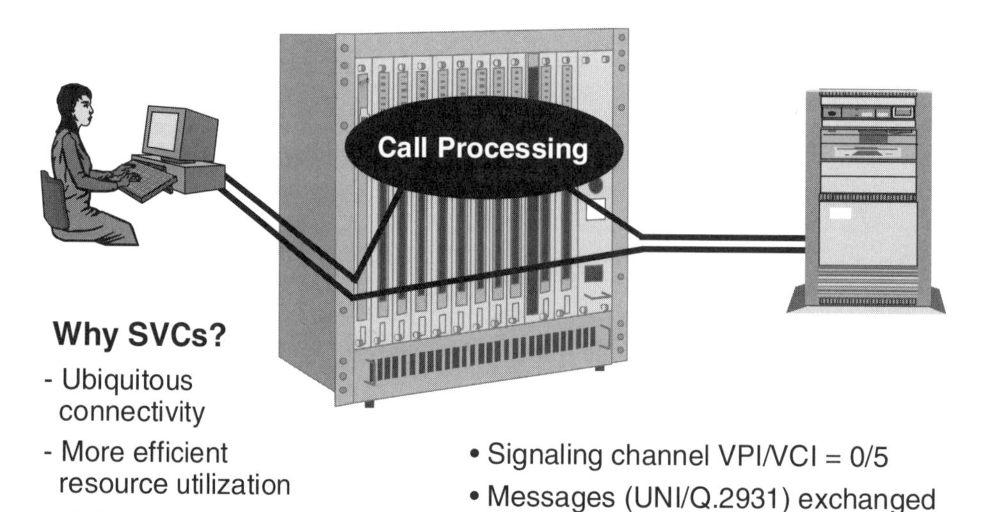

Switched Virtual Connections (SVC)
Fig. 9-3

connection type, bandwidth and QoS; and VPI/VCI values are initialized and assigned from attached devices and all intermediate ATM devices. Once this is accomplished and the terminating device signals it is ready to receive information, a **Connection–established** message is communicated to the source that connectivity exists and information can flow. Each network element acknowledges the Connect message by sending a **Connect Ack** message, completing the connection establishment phase. Figure 9-4 illustrates a simple call setup of a connection involving the use of UNI access signaling protocol from the user and network signaling protocol between network elements interconnected with the NNI. This makes ATM a connection-oriented communication technique. When the communication session or information transfer is complete, a release signaling message is sent, thereby ending the connection. A more complete discussion and treatment of ATM Broadband signaling protocols is contained in Reference 7.

Connection setup performance requirements for a switch to execute the functions necessary in establishing a connection through a switch is specified in *Broadband Access Signaling Generic Requirement* GR-111-CORE[8] as a maximum of 200 milliseconds. However, setup times very significantly depend on whether the connection is a simple ATM connection or a complex call with additional supplementary services to be provided. Since this 200-millisecond benchmark was established, ATM switching and call processing technology has progressed to the point that simple SVC connection set up in a range of 8 milliseconds to 50 milliseconds. This becomes a point of competitive differentiation among the vendors.

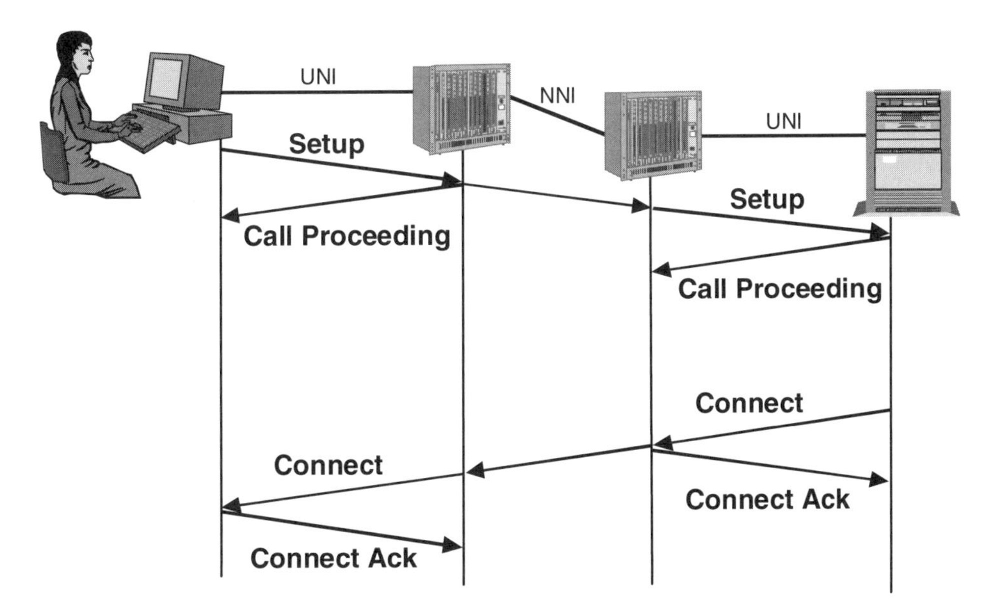

Simple Call Setup of a Connection
Fig. 9-4

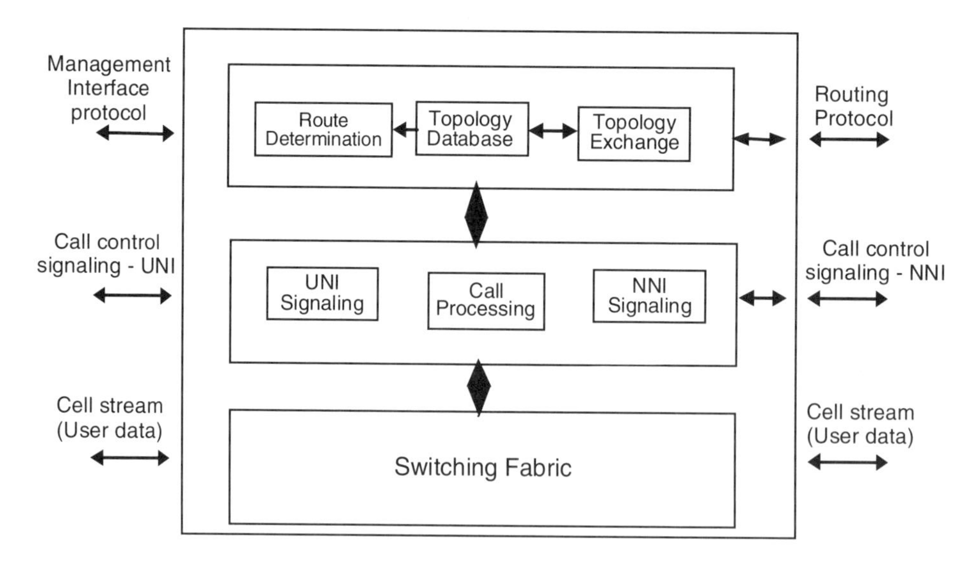

Functional Block Diagram of Broadband Switching System
Fig. 9-5

Once the connection setup is complete, cells can flow. The worst case ATM switch delay from ingress port to egress port is 150 microseconds. When cells arrive at a switch node, the only processing that is required is on the VPI and VCI values which determine the output link to forward the cell. Figure 9-5 contains a functional block diagram of a broadband switching system.

Throughout connection setup and while user information is flowing, tests are performed to ensure reliable service. These tests include:

- Verifying continuity utilizing the ATM Operation Administration and Maintenance (OA&M) cell capability
- Verifying resource availability, monitoring of traffic and QoS

These additional ATM mechanisms and capabilities contribute to more reliable networks that are easier to maintain and operate.

The discussion above highlights the advantage of taking a little more time to signal and establish an end-to-end connection. This contrasts with the IP approach of simply sending a packet of information. In the IP approach, each IP packet must contain source and destination address information. Each router has to lookup and process every IP packet before determining a route. [Standards are under development in the IETF to shortcut some of the IP routing procedures and to enable cut-through switching, and facilitate IP switching based on QoS like constraints, such as in Multi Protocol Label Switching (MPLS).]

Broadband user network signaling stack software has to be loaded or implemented in the user's equipment and in the line side of the network peer. The User Signaling Entity in the broadband terminal communicates with the Call Processing Entity in the network switch. UNI signaling consists of a three-layer protocol stack based on ITU-T Recommendations. It is based on the signaling protocols developed for narrowband ISDN but is adapted for ATM

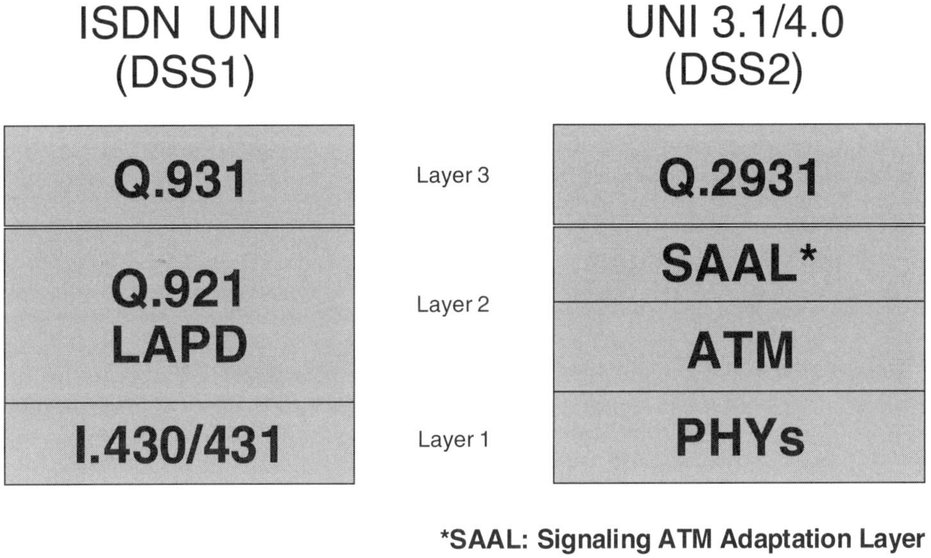

ISDN UNI (DSS1)

UNI 3.1/4.0 (DSS2)

ISDN UNI (DSS1)		UNI 3.1/4.0 (DSS2)
Q.931	Layer 3	Q.2931
Q.921 LAPD	Layer 2	SAAL* / ATM
I.430/431	Layer 1	PHYs

***SAAL: Signaling ATM Adaptation Layer**

Narrowband and Broadband User-Network Signaling Protocol Stack
Fig. 9-6

technology. This set of protocols is referred to as Digital Subscriber Signaling (DSS). DSS 1 indicates the narrowband ISDN user signaling, while DSS 2 refers to Broadband user signaling, Figure 9-6. The ATM Forum has extended the ITU Q.2931 Recommendation and is included in the UNI Signaling version 4.0 specification.

9.2 PRIVATE NETWORK-TO-NETWORK INTERFACE (PNNI)

PNNI was developed to enable large private/enterprise networks based on ATM to support all the services already defined and/or adapted to ATM, thus facilitating new, higher performance corporate networks. This required the development of a switch-to-switch signaling mechanism. The decision was to base PNNI signaling on the UNI v4.0 access signaling specification version, but it incorporates two significant differences. First, the UNI signaling protocol is not a symmetrical protocol. The Call Processing Entity on the network side of the connection is responsible for allocating resources and connections. The end-point User Signaling Entity communicates to the network the characteristics of the connection, and the network instructs the terminal equipment which VPI/VCI to use for that connection. The network side of the connection is in control. In the case of switch-to-switch connections for PNNI signaling, either end of the connection contains a Call Processing Entity which may request and allocate a VPI/VCI. This could result in contention with VPI/VCI assignments and resource allocations if each side simultaneously makes similar decisions. So the protocol procedures were modified for symmetrical peer-to-peer operation by adding contention resolution procedures.

The second area where PNNI differs from UNI signaling is the addition of a dynamic state routing protocol. Rather than manually defining and establishing static routing tables for each switch, the objective was to adapt dynamic routing protocols used by the Internet for ATM SVCs. The hierarchical topology state source routing protocol builds on top of UNI signaling. This can simplify network configuration management and routing table set up. It also allows more effective utilization of network resources. This is discussed further in Section 12.

9.3 NETWORK SIGNALING

Most public telecommunications network-to-network signaling is based on Common Channel Signaling (CCS) protocol, which evolved from the 1970s. CCS carries signaling information from switch to switch, generally over a separate signaling network dedicated for this purpose. Just as ATM broadband user-to-network signaling was based on extensions to narrowband protocols, broadband network signaling is based on extending the narrowband network signaling protocols, Figure 9-7. The lower three CCS protocol layers are called Message Transfer Part (MTP), with the number indicating the layer. MTP is responsible for the reliable transfer of signaling information over duplicated and alternate routed links across the CCS network to the destination. In addition, functions allow the MTP layers to respond and overcome system failures that would affect the transfer of signaling information.

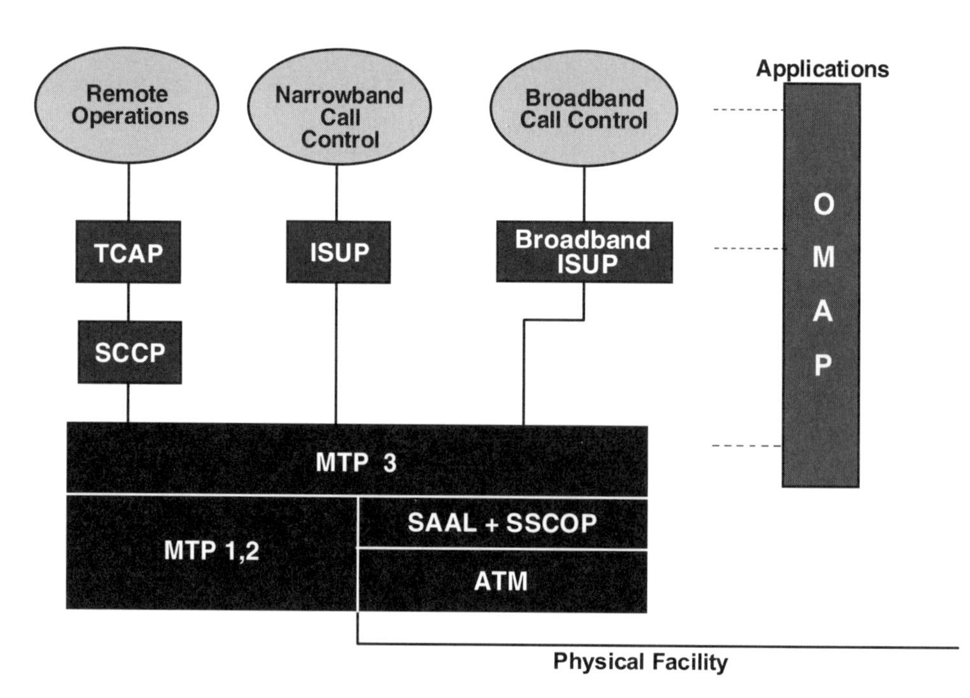

Common Channel Signaling (CCS) Protocol Stack
Fig. 9-7

MTP 1 specifies the physical, electrical, and functional characteristics of the signaling link. MTP 2 defines the functions and procedures for the transfer of messages over a link in an error-free and reliable manner. Messages are transmitted using signal units that follow the framing conventions of the high-level data link control (HDLC) procedures. MTP 3 defines networking (transport) functions common to all links in the end-to-end connection. MTP 3 is a connectionless protocol allowing distribution of messages to multiple MTP users and maintaining correct message sequencing. The collection of these three layers make MTP responsible for the physical movement of signaling information. The lower three layers are common to all CCS network signaling functions.

Above the MTP layers, the CCS protocol stack splits into two branches, one for call control and the other identified as remote operations. The call control branch introduces a protocol layer called ISDN User Part (ISUP). ISUP defines the protocol to control inter-exchange connections for calls that provide switched services and user facilities for voice and data applications in ISDN. ISUP defines the signaling messages, IEs, encoding, and cross-office performance. ISUP is generally the network-side equivalent to the Q.931 layer of user access signaling protocol which enables end-to-end connection establishment of basic bearer services and supplementary services.

The Broadband ISDN User Part (BISUP) encompasses the signaling functions required to provide switched Broadband services and user facilities for voice, video and data applications. In addition, by building on ISUP, BISUP supports narrowband services and features, and enables easy interworking between narrowband services on an ATM broadband network and with userscconnected to the existing network.

It is important to note that BISUP operates over the existing MTP layers without change. This allows the easy migration of ATM broadband service without major upgrades to the existing CCS network. However, with increasing use of network intelligence and number-driven services requiring access to database information, the CCS network is being adapted to use ATM transport for higher performance signaling. To achieve this, MTP 2 layer is replaced with the combination of sub-layers called Signaling ATM Adaptation Layer (SAAL) plus the Service Specific Connection Oriented Protocol (SSCOP). These two sub-layers provide the equivalent MTP 2 layer functions. All other layers of the CCS stack remain unchanged independent of narrowband or ATM broadband facilities.

Intelligent network services that are based on database queries utilize the Remote Operations branch of CCS signaling protocol stack. Signaling Connection Contro! Part (SCCP) provides the means to (1) control logical signaling connections in a CCS network, and (2) routing both connections and datagram messages across the CCS network using logical addresses such as an 800 number. SCCP also supports duplication of database functions to provide additional availability for essential services.

The Transaction Capabilities Application Part (TCAP) provides the means to exchange operations and replies via a dialog. TCAP provides the mechanism to establish non-circuit-related communications between two points in the signaling network, such as a switch and a service control point. The combination of SCCP and TCAP are essential for intelligent network services.

The CCS signaling network is a packet switched network with dedicated links enabling signaling information from a large number of different users to be multiplexed. CCS operates in different modes: associated, non-associated, and quasi-associated signaling. (See Figure 9-1: Realms of Signaling.)

1. Associated: In this mode, signaling messages regarding a particular signaling relation between two adjacent points are conveyed over a direct link interconnecting these signaling points.
2. Non-associated: In this mode, messages relating to a particular signaling relation are carried over two or more links in tandem and pass through one or more signaling points other than those that are the origin and the destination of the message.
3. Quasi-associated: This mode is a subset of the non-associated mode. The signaling path taken by the messages is predetermined and switched through Signal Transfer Points (STP) before reaching a distant switch.

Implementation of fully associated CCS would require point-to-point signaling links between any two switches. As the number of switches in the network grows, and each switch needs to have a dedicated associated mode pair of links to every other switch in the network for a full mesh configuration, the number of links required grows by a logarithmic factor. For small networks and during early ATM deployments, the associated mode signaling may provide lower connection setup delay and costs. The network, however, cannot grow very large before it is no longer practical to interconnect switches using the associated mode of signaling.

Non-associated signaling, also referred to as quasi-associated mode of signaling, is implemented over a network using STPs operating as connectionless packet switches, see Figure 9-1. This topology approach eliminates the need for a large number of point-to-point links interconnecting between every switch. Within an area code area, it is typical to have hundreds of switches. If each were required to have duplicated (for reliability) associated signaling links connecting to every switch within the area code in a mesh topology, all the interfaces would be dedicated to signaling and would be cost prohibitive.

Signals are communicated between switches over two or more links and one or more STPs in tandem. The function of the STP is to relay signaling packet messages between the various constituent links. The STP provides protocol link level 2 functions (as defined in the OSI reference model) responsible for error detection/correction, and performs protocol level functions 3 for network routing, along with traffic management and operations. This network signaling architecture is more efficient because it eliminates the need for a switch to have signaling links to all interconnecting switches. A more complete treatment of signaling protocols can be found in Reference 7.

Operations Maintenance and Administration Part (OMAP) provides the management capabilities to support individual layers and the CCS system network.

Finally, PNNI signaling was developed for large-scale, private switch-to-switch signaling and control can be used as an alternative to CCS operating in the associate mode. The advantages include the avoidance of connecting the ATM broadband network into the existing CCS network or avoiding entirely the need for CCS network element deployment for new service providers/operators, thus reducing initial capital investment. In addition, routing tables in the existing circuit-switched network must be statically defined. With the rapid network growth, PNNIs ability to dynamically route simplifies configuration management. However, two limitations also need to be considered. First, while PNNI will route dynamically, care must be taken in establishing the hierarchical layout of the network. Secondly, the PNNI signaling protocol does not have any transaction capabilities. Consequently, services such as 800 numbers, mobility, and others previously discussed cannot be supported at this time.

The use of PNNI will be discussed further in Section 12, Public and Private Network Architectures and Interconnect.

10 TRAFFIC MANAGEMENT

The multi-service demands of broadband services introduce some new challenges. The service characteristics of voice, data, and video are significantly different. These services require continuous streams that run the gamut from real-time-dependent information to random bursts of information and operate over a wide range of speeds (Kbps to Gbps). There are significant uncertainties when it comes to predicting the characteristics of future applications and multimedia services for broadband. This uncertainty combined with unknown demand for these broadband services has a significant impact on the design and evolution of backbone networks and the technology basis for the network.

The design of broadband networks must consider a number of criteria. First, to satisfy the unknown growth pattern in future service demands, it is important to have a robust network technology design that can be modified easily in response to changes in demand for a particular communications service. For example, it is not possible to anticipate the traffic needs five years from now. Within a five-year period, significant new applications and/or technologies emerge or are invented, such as the Web Internet. Second, the network must be able to handle vastly different types of traffic ranging from very low-speed data to voice to full-motion video. Third, depending on the demand for the broadband services, a network design must be capable of providing a migration strategy from the existing environment to succeeding generations of high-speed packet-switching and fiber optic transmission which are used in connection with the delivery of broadband services.

If these challenges were not enough, let us now consider other aspects that have led to the development of specific ATM traffic management capabilities. QoS is a multi dimensional problem with delay, latency, and cell loss requirements. Multimedia applications utilizing one or more media types (voice, video, and data) also have different QoS needs which must be supported simultaneously as illustrated in Figure 10-1. Carriers/operators want to maximize the network resources and fill the trunks to the highest capacity possible while still guaranteeing end-to-end QoS.

As ATM cells traverse the network, a number of things can happen to them. One network element may be slower

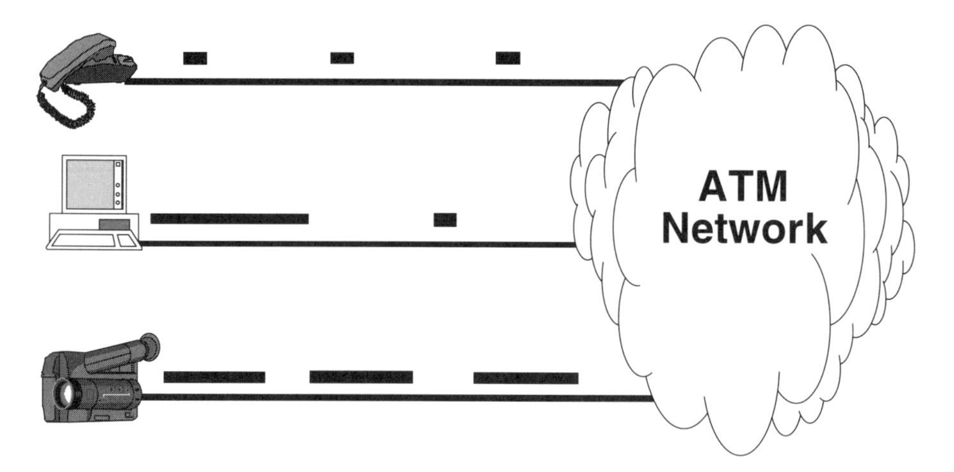

Different applications need different QoS and have different characteristics.

Traffic Management Challenge
Fig. 10-1

than another and introduces some amount of delay. Over the course of the network, with information coming in from different sources and network elements, "clumping" of cells can occur leading to cell loss. In addition, ATM switching and multiplexing will cause cell delay, cell delay variation, and potentially lost cells. To address these challenges, the approach taken in developing the traffic management mechanisms is illustrated in Figure 10-2. The user communicates to the network what is needed. For PVCs, this occurs at subscription time, but the future network must be designed to offer and support on-demand or SVC services. In the case of SVCs, the signaling protocol communicates the characteristics of the desired connection to the network. Note: The user does not need to know about the details of ATM traffic management, signaling protocols or the type of information necessary to establish a connection. The application software in the terminal device, upon initiation will communicate automatically the characteristics of the connection through an Application Program Interface (API). The network determines what is available based on the traffic parameters and QoS and establishes the connection. Once the connection is established, the network must provide some type of policing mechanism to ensure that the bandwidth or other contracted attributes are not exceeded and do not interfere with other subscribers.

ATM traffic management has six components:

- Resource Management (RM) control
- Connection Admission Control (CAC)
- Traffic Shaping (TS)
- Cell Loss Priority (CLP) control
- Explicit Forward Congestion Indication (EFCI)
- Usage Parameter Control (UPC)

Each component will be briefly described.

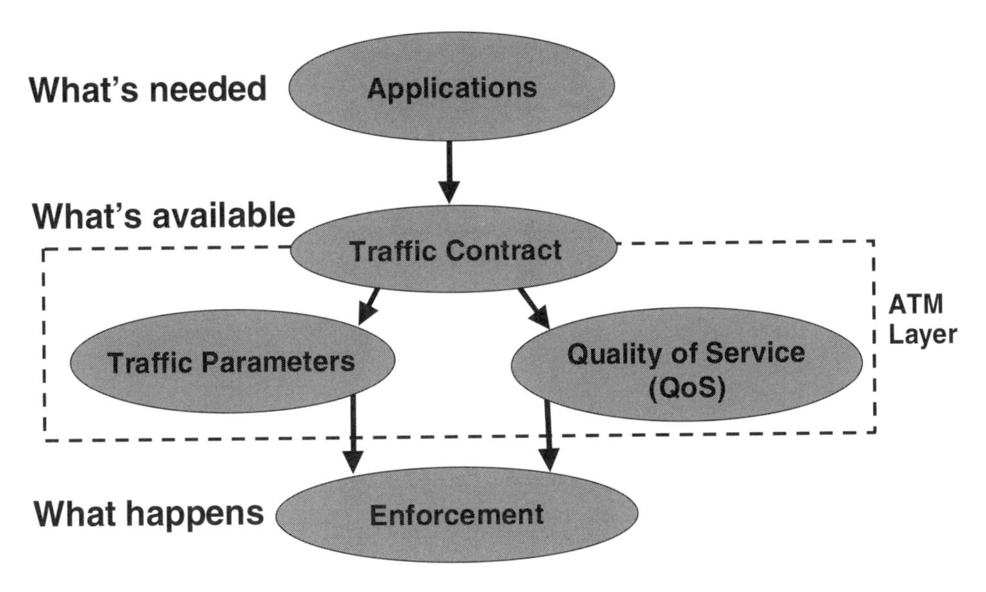

Approach to ATM Traffic Management
Fig. 10-2

Resource Management is a technique used by the network provider. Network management support systems are used to provision certain resources in each network element within the network. They also predefine static routing tables to maximize utilization of network capacity and avoid congestion. (Details and guidelines on network traffic engineering go beyond scope of this book.) One common approach to simplify the network traffic engineering problems is to segregate the different types of traffic onto separate connections. For example, all the voice traffic, which has similar characteristics and QoS needs, could be bundled onto one Virtual Path. Video could be carried over a different VP or interface, and data on another. The only limitation with this method is that it is manually provisioned, and responding to changes in near real time is not possible. More automated mechanisms such as Connection Admission Control are necessary and provided with SVCs. PNNI goes further with its dynamic routing capabilities based on traffic management and link state information. However, for PVCs and for initial traffic design layout of a backbone network, the static routed approach to resource management is a good place to start.

Connection Admission Control (CAC) is a process that uses traffic descriptors, to define the characteristics of the requested connection and maintain knowledge of the current network traffic capacity and committed load to other users. CAC determines whether or not the new connection request can be honored without degrading the QoS of existing services. CAC utilizes standardized traffic descriptors carried in standardized signaling protocols. The CAC technique is not subject to standardization because it is internal to a switch and dependent on service provider/operator policies and practices. This provides a basis for competition.

Traffic shaping refers to the function provided on a hardware interface. The ATM cell traffic is "monitored" (sometimes also called policing) to smooth the traffic and avoid "clumps" that might result in cell loss due to delay variation. When the traffic shaping function is applied to the network, it spaces cells so as not to exceed the Peak Cell Rate (PCR). PCR is one of the traffic parameters negotiated with connection establishment. Traffic shaping can also be performed at the egress of the user terminal equipment. When applied by the user, it helps reduce burstiness as well as ensure that the PCR is not violated.

Cell Loss Priority (CLP) Control, frequently referred to as "tagging" or selective discard, utilizes the 1-bit ATM cell header field called CLP. The CLP bit indicates the relative cell loss priority and assists the network during times of congestion in selectively discarding cells and minimizing degradation of network service to all users. If the CLP is set to 1 (CLP = 1), the cell is subject to discard, depending on network conditions and network/operator policies. If CLP is not set (CLP = 0), the cell has a higher priority. CLP = 0 are low loss priority cells. CLP = 1 are high(er) loss priority cells. CLP = 1 traffic has higher CLR (Cell Loss Ratio) than CLP = 0.

Tagging is a procedure where cells violating traffic contract are converted from CLP = 0 to CLP = 1. The carrier/operator may tag or discard violating cells. Tagging marks non-conforming cells by changing the CLP from 0 to CLP of 1 thereby giving those cells a higher loss priority. The objective of tagging is to allow those cells to continue to be transported through the network if the resources are available. However, if congestion occurs, CLP = 1 cells are the first to be discarded. Carriers/operators can choose to use this tagging procedure or simply discard violating cells. The approach chosen is based on carrier/operator policy decision.

Explicit Forward Congestion Indication (EFCI) is a mechanism that allows one node that is experiencing congestion to communicate this in all cells leaving the node. EFCI uses the Payload Type Indicator (PTI) field for this purpose. EFCI allows a receiver to notify the transmitter of congestion. The transmitter can take corrective action and slow down to reduce the probability that cells may be lost. However, because the mechanism operates in the forward direction, an intermediate switch that may be experiencing congestion must wait for the forward transmission of the EFCI cells to the destination; the destination then must respond to the transmitting source before the transmitter can take action. This "permissive" operation, depending on round trip delay, may not help much. However, when combined with other traffic management techniques, including CLP or alternate routing, it may be sufficient. The effectiveness also is contingent on the user doing the right thing. One inducement is to develop charging or billing strategies that service providers/operators can levee to encourage the correct behavior.

The Usage Parameter Control (UPC) mechanism, like traffic shaping, is primarily hardware controlled and policed. Nine parameters have been defined and standardized to allow users to communicate the characteristics of their connection and to allow SVC interoperability between carriers/operators for end-to-end connections. These parameters are identified below, but the range of values and additional details are not addressed. (Most users will never know anything about these parameters because application software developers make it transparent.) The parameters are:

- PCR = Peak Cell Rate
- SCR = Sustainable Cell Rate
- MCR = Minimum Cell Rate
- CLR = Cell Loss Ratio
- MBS = Maximum Burst Size a function of (PCR, CDVT, SCR, BT)
- BT = Burst Tolerance
- CTD = Cell Transfer Delay
- CDV = Cell Delay Variation
- CDVT = Cell Delay Variation Tolerance

These six ATM traffic management components provide effective network traffic management strategies and generic cell rate algorithms. These are left for more specialized reading.

The ***Available Bit Rate (ABR)*** service was born from experiences of Internet Protocol (IP) and uses some of the parameters described above. IP is a Variable Bit Rate (VBR) packet data protocol that assumes all communications get the same level of service, and it is a "best effort" service. There is no guarantee that the IP packet will reach its destination. If more reliable packet data transfer is needed, that the end user must employ additional higher layer protocols that can detect and retransmit packets if there are errors or if packets are lost. With IP traffic growth and the dynamic and unpredictable nature of IP best effort service, a few concerns emerge. These include:

- IP/Internet users want to get as much available bandwidth as possible on an as-needed basis. The IP user assumes it will cost less than a regular subscription or Service Level Agreement (SLA) because, if it is available, the bandwidth is not being used by other customers. Carriers/operators can accept the IP packet flow if resources are available but can take away those resources for other higher QoS customers when needed.
- A minimum amount of bandwidth may be necessary to stay "alive."
- With very unpredictable traffic swings, better flow control is important and enables some traffic management components described previously to avoid congestion collapse in data networks.

ABR service was designed for delay-tolerant and cell-loss-intolerant data applications. While ABR is patterned after IP best effort service, it introduces a dynamic allowable cell rate-based network feedback mechanism and provides for a minimum cell rate. This assures some minimum level of bandwidth or throughput for those IP based applications where it is critical. ABR provides a higher class of service than best effort. The ABR mechanism required the definition of a *Resource Management* (RM) cell, illustrated in Figure 10-3. The RM cell is identified by the ATM PTI field in the ATM cell header. In this way, even when a specific connection is congested, the RM cell can be easily communicated.

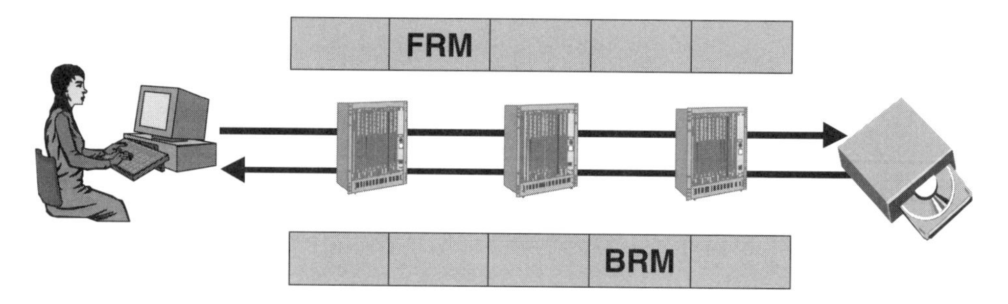

- Forward Resource Management (FRM) cell lists PCR, MCR, etc. requested by terminal. Network equipment modifies (reduces) if unacceptable.
- The Backward Resource Management (BRM) cell informs the network equipment to commit to selected rate(s).
- Virtual Sources (VS) and Virtual Destinations (VD) can be added to reduce round trip delay.

Available Bit Rate (ABR) Service
Fig. 10-3

ABR has four modes of operation:

- *Explicit Forward Congestion Indication* (EFCI) Binary Rate: EFCI is a simple end-to-end flow control using the EFCI PTI field in data cells.
- *Relative Rate Binary Operation* (RRBO): RRBO uses the Resource Management (RM) cell that was defined for ABR service. Binary fields in the RM cell request increasing or decreasing the bandwidth.
- *Explicit Rate Operation* (ERO): This mode uses the RM cell, but explicitly communicates the rate at which the source wants to operate.
- *Virtual Source/Virtual Destination* (VS/VD): VS/VD is a technique that allows the feedback loop to be shortened. Unlike EFCI, which must first be transmitted to the destination before turning it around and sending it back to the transmitter or source, VS/VD allows the ABR to work end-to-end, subnet-to-subnet, or segment-by-segment.

It should be noted that during connection setup, switches can negotiate the ABR parameters down through signaling.

The traffic management components and techniques discussed in this section enable ATM to guarantee QoS, rates, delay, and cell loss, so that large-scale multi-service networks can be deployed and user needs and expectations regarding services from the network, private or public can be met. ATM traffic management also is a critical element in realizing carrier/operator interoperability. Finally, combined with signaling, the manually intensive effort of PVC service provisioning or negotiating SLAs essentially is automated. The benefit of automating PVC service or negotiating SLAs becomes even more significant when the number of subscribers or attached devices goes into the thousands or millions.

Guaranteed Frame Rate (GFR) Service is the most recent traffic management service that incrementally provides a high level of service. GFR is intended to support non-real-time applications. It is designed for applications that require a minimum rate guarantee and can benefit from dynamically accessing additional bandwidth available in the network. but it is not critical to the continued effective operation of the application. It does not require that the application adhere to a flow control protocol. The service guarantee is based on AAL5 PDUs (frames). Another feature of GFR service is that if a cell has an error or one cell making up the PDU or frame is lost, the network will discard all the cells making up that PDU rather than continuing to transmit cells that would be useless when received at the destination. This avoids sending cells during congestion that may be contributing to network overload ultimately resulting in ATM cells being arbitrarily discarded under congestion. This also alleviates end user terminal equipment having to reassemble the PDU from the received cells, determining whether an error had occurred, discarding all the cells making up that errored PDU, and initiating retransmission.

When establishing the GFR service connection, the end user system equipment specifies a PCR and a Minimum Cell Rate (MCR) which is defined along with a Maximum Burst Size (MBS) and a Maximum Frame Size (MFS). The GFR service can be specified with an MCR of zero. The user may send cells at a rate up to the PCR, but the network only commits to carry cells in complete frames at the MCR. Traffic beyond the MCR will be delivered within the limits of available resources.

11 MANAGEMENT PLANE

The management plane includes the management layer functions and the plane management functions as shown in the three dimensional Protocol Reference Model contained in Figure 4-2. Layer management interfaces with the individual layers of the model, such as the physical layer, ATM layer, AAL, and higher layers. Plane management is responsible for coordination across the layers and planes in support of the user and control planes through layer management facilities. Plane management ensures that everything is functioning properly on a system level.

11.1 LAYER MANAGEMENT

A hierarchy of five flows associated with the OA&M functions has been defined in ITU-T Recommendation I.610. It starts with the SONET/SDH physical layer through the ATM layer. These flows are frequently identified as F1-to- F5. Flows F1, F2, and F3 apply to the SONET/SDH physical layer. The flows are referred to Section, Line, and Transmission Path respectively or simply Path level. The functions provided by each of these levels is discussed further in Part 2.

The ATM layer introduces two additional levels of OA&M flows, Virtual Path flows (F4) and Virtual Channel flows (F5) between connection end points, Figure 11-1. The F4 and F5

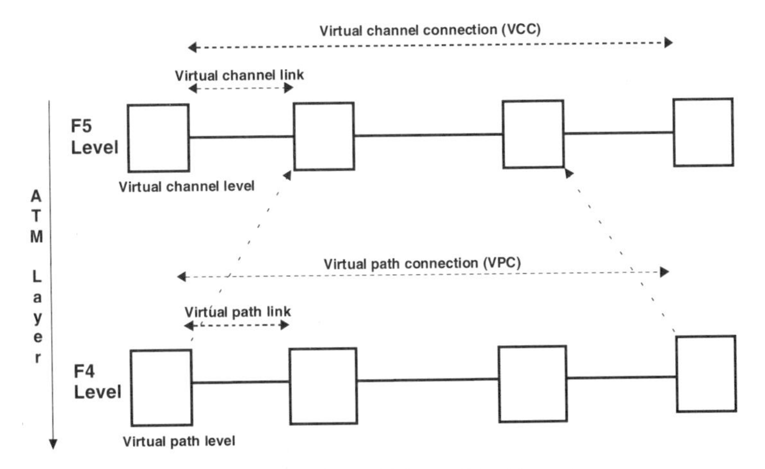

ATM Layer OA&M Hierarchical Levels
Fig. 11-1

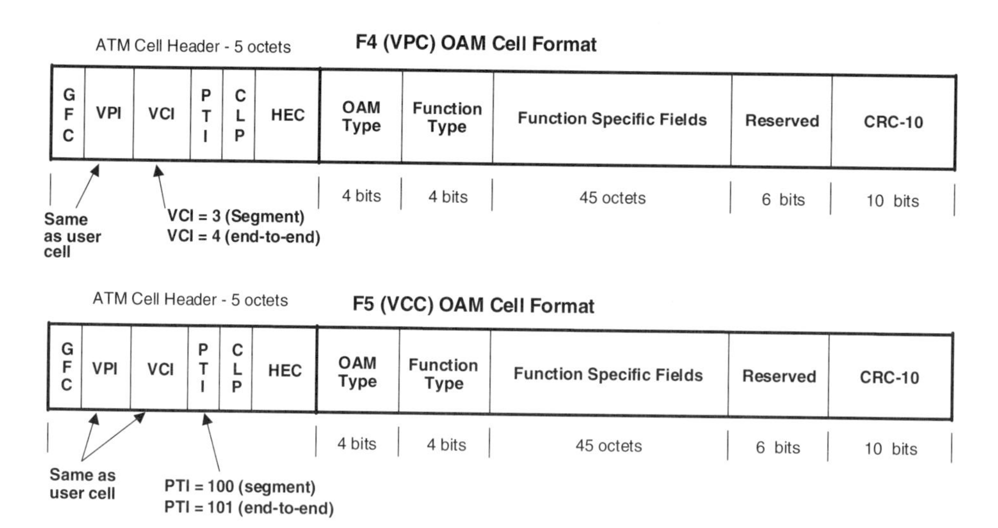

ATM OA&M Cell Type and Format
Fig. 11-2

flows may occur across one or more interconnected VC or VP links that segment OA&M flows.

VP flows (F4) utilize different Virtual Channel Identifiers to identify whether the flow is end-to-end (VCI = 3) or segmented (VCI = 4), Figure 11-2. As a VP traverses a network, an incoming VP bundle may be mapped to a different outgoing VP by cross connect type equipment or through switches. By identifying whether an ATM OA&M cell is intended to be end-to-end or for one segment enables network fault isolation.

VC flows (F5) are specific to a VCI and predefined VCIs cannot be used to identify whether the VC OA&M F5 flow is intended to be end-to-end for that Virtual Channel Connection (VCC) or intended just for that Virtual Channel Link or segment. To resolve this and enable troubles or faults to be located and isolated, the Payload Type Identifier (PTI) field in the ATM cell header is used to differentiate between end-to-end (PTI = 100) and segment (PTI = 101) flows in a VCC. See Table 2-1, page 11.

Table 11-1 summarizes the OA&M type and function type fields in the OA&M cells shown in Figure 11-2. The three OA&M types are fault management, performance management, and activation/deactivation. Each OA&M type has additional function types with code points identified in Table 11-1.

There are four fault management OA&M types: Alarm Indication Signal (AIS), Remote Defect Indication (DRI) also known as Far End Reporting Failure (FERF), Continuity Check, and Loopback function types. For the performance management OA&M type, there are forward monitoring and backward reporting types, and a third type, which is a combination of these two, called monitoring and reporting. The third OA&M type defines activation and deactivation functions for performance management and continuity check. A significant number of

OAM Type		Function Type	
Fault Management	0001	AIS	0000
	0001	RDI/FERF	0001
	0001	Continuity Check	0100
	0001	Loopback	1000
Performance Management	0010	Forward Monitoring	0000
	0010	Backward Reporting	0010
	0010	Monitoring & Reporting	0010
Activation/Deactivation	1000	Performance Monitoring	0000
	1000	Continuity Check	0001

OA&M Types and Function Types
Table 11-1

unassigned code points are contained in the OA&M cell function types. Use of these code points will evolve over time as new needs are identified and standardized.

11.2 PLANE MANAGEMENT

Plane management is based on Operations, Administration, Maintenance and Provisioning (OAM&P) functions necessary in running a network. These functions are generically defined in the ITU-T and OSI standardized Common Management Information Service Elements (CMISE) and associated Common Management Information Protocol (CMIP). These functions and protocols are used in the Telecommunications Management Network (TMN) architecture.[11] The following describes the interfaces and specifications developed for ATM equipment.

To manage the ATM technology in a system or on a network element basis, several network management interfaces have been developed or are undergoing development. These interfaces are referred to here as M1 through M5 and illustrated in Figure 11-3. The network management functions include Performance Management (PM), Fault Management (FM), and Configuration Management (CM).

The User Network Interface (UNI) predefines a VCI for the network management channel between the end user and the attached network. This management channel is called the Integrated Local Management Interface (ILMI) [28] and is identified as the M1 interface, Figure 11-4. The ILMI communicates over a predefined VPI/VCI and uses the SNMP (Simple Network Management

[11] Additional details are available in ITU-T Recommendation M.3010—"Principles for Telecommunications Management Network," 1992, and in B. Hebrawi, *OSI Upper Layer Standards and PRactices*, McGraw-Hill, 1992.

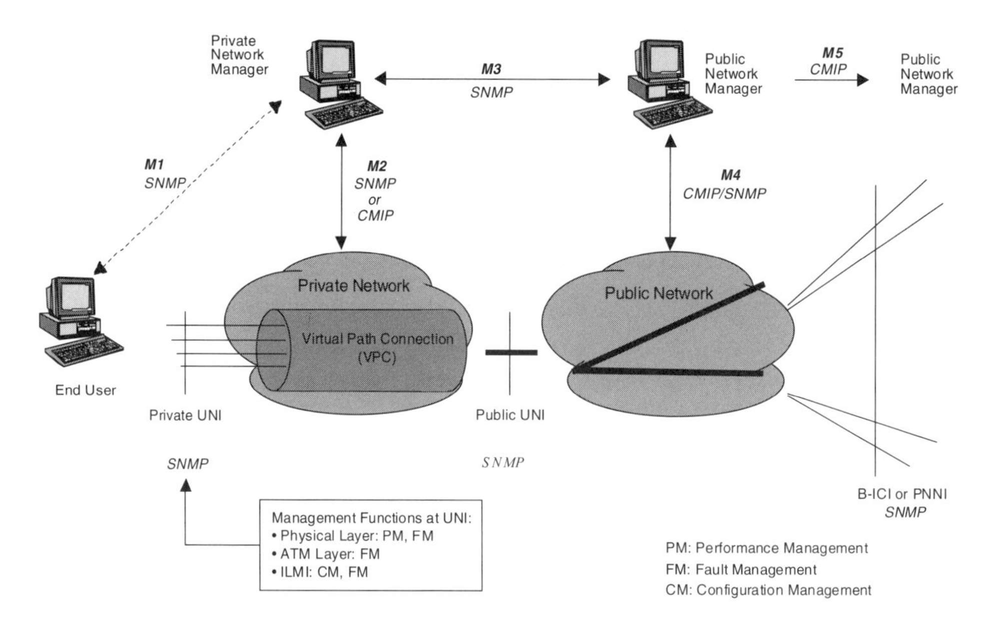

Network Management Reference Configuration
Fig. 11-3

Protocol) extended and adapted to ATM. The ILMI provides users with several key features. They include the customer's ability to initiate requests for service reconfiguration or changes in subscription parameters, obtain usage information, and perform address registration. Because users do not connect directly to the network manager, the M1 connection shown in Figure 11-3 is a logical connection, not a physical connection.

While not all carriers/operators support these features at this time, in the future it will be much easier for users to move their own terminal devices, plug them in at a new office location, and obtain services. This is achieved by using

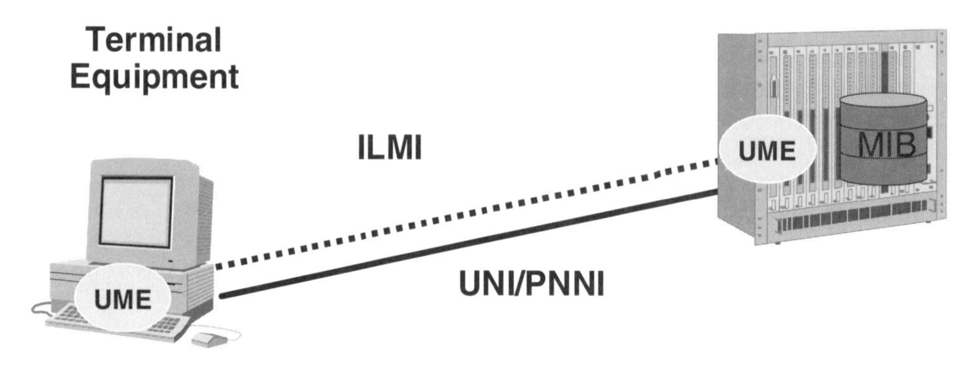

Integrated Local Management Interface (ILMI)
Fig. 11-4

ILMI protocol procedures to determine the port address for that location, service parameters (bandwidth, number of VPIs/VCIs allowed on that interface, classes of service, etc.) that may be predefined in the case of business environment. The ILMI/M1-to-NMS (Network Management System) communications provide the customer a view of the carrier/operator network.

The M2 interface provides the interface between a private Network Management System (NMS) and the private ATM network elements that it controls. Specifications for the M2 interface have not been developed and are vendor proprietary.

The M3 interface links a customer's private NMS to the public carrier/operator NMS. This enables customer requests for service reconfiguration, customer and carrier/operator exchange of fault and configuration data. This will greatly help to reduce the time required in identifying, isolating and resolving failures.

Similar to the M2 interface, the M4 interface is located between the public carrier/operator NMS and the ATM Network Element (NE) it controls. The typical NMSs developed for public networks can manage a large number of NEs such as switches, routers, access multiplexers and cross-connects. The NMS through the M4 interface provides the connecting NE with critical information regarding user subscription information and with the services that the user is eligible to receive, provides fault and configuration management across the entire network that it is managing, and provides a window into network performance. The information gathered and tracked through M4 communications also provides valuable information and input for network capacity planning. The M4 and M2 interfaces use either SNMP or CMIP.

The M5 interface will enable communications and information exchange between carrier/operator Network Management Systems. For example, interconnecting carriers/operators will be able to exchange network fault and configuration management information, usage information in support of billing, and service requests for communications going between carrier customers. Development of the M5 interface is currently underway.

12 PUBLIC AND PRIVATE NETWORK ARCHITECTURES AND INTERCONNECT

ATM enables a multiservice networking platform supporting a variety of services over shared interfaces and backbone. However, with embedded investment that private networks and public networks have in existing communication technologies, it is not practical to replace this investment and the applications with all new equipment overnight. Most have adopted an incremental investment strategy for migrating their networks made up of a variety of existing service-specific technologies (Frame Relay, X.25, IP, Hybrid Fiber-Coax, etc.), with each built and managed separately, Figure 12-1. Users also have embedded investments and legacy systems that they want to continue to use independently of the technology basis service providers/operators choose for their networks. This places additional requirements on ATM-based Network Elements and interface specification development to support existing user and network equipment containing non-ATM service specific interfaces and applications. Examples include:

- Frame Relay interfaces using High-level Data Link Control (HDLC) protocols on DS1 facilities
- SMDS Subscriber Network Interfaces (SNI) using the Institute of Electrical and Electronics Engineers (IEEE) 802.6 protocol standard
- Emulated T1/E1, T3/E3 interfaces

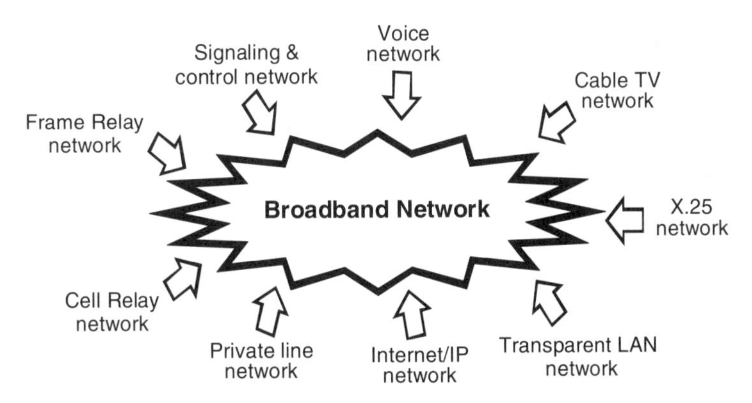

Multiple Service-specific Networks Evolve Toward a Broadband Network
Fig. 12-1

- Interworking functions to connect B-ISDN to Narrowband ISDN (N-ISDN)
- Ethernet and Token Ring interfaces
- IP/Internet router-based networks

The support of applications and application environments, including Local Area Network (LAN) protocols, are discussed further in Part 3 of this book. A fundamental consideration for the interconnection of networks and embedded equipment begins at the physical layer, and whether or not the existing "wiring" can be used with the new equipment. If not, it adds significantly to the cost of upgrading and migrating to that technology let alone the inconvenience and delay to install/pull new wires or fiber cables.

12.1 NO RE-WIRING NECESSARY

ATM works with any Physical Layer technology. A comprehensive set of Physical (PHY) Layer specifications have been developed to allow the use of existing cabling structures, optical fiber, and wireless technology as we move towards global information infrastructure. Approved Physical Layer specifications include Unshielded Twisted Pair (UTP) wiring for Category 3 and Category 5 (usually referred to as UTP3 and UTP5) commonly used for inside wiring. UTP3 and UTP5 support physical interface rates of 25 Mbps, 51 Mbps, and 155 Mbps. Network Interface Cards (NICs) are available that plug into existing PCs or workstations to support ATM connections back to a small ATM switch or hub without any wiring changes. Physical interface specifications have also been developed for multi-mode and single-mode fiber at 100 Mbps, 155 Mbps, and fiber driven by LEDs or by lasers and higher speed interfaces of 622 Mbps and 2.4 Gbps. Work on 10 Gbps interfaces. (Exploring both SDH and Wave Division Multiplexing (WDM) approaches and wireless ATM is underway.) The set of PHY specifications is too numerous to reference in this text. Readers can download the information from www.atmforum.com/atmforum/specs/approval.html.

These specifications are one important aspect facilitating the migration to ATM solutions and infrastructures capable of supporting multimedia. This is important because users can select the interfaces that best meet their needs and match the embedded wiring already in place. Users do not need to install new wires/cables to utilize higher speed ATM. ATM is also compatible with the transmission hierarchies that service providers and network operators have deployed around the world.

12.2 WIDE AREA NETWORK INFRASTRUCTURE AND INTERWORKING

The objective is to have end-to-end connectivity between end users with no regard to the type of network to which those end users are connected. The interface specification for interworking between private ATM networks is based on Private Network Node Interface (PNNI) or ATM Inter Network Interface (AINI). The interface specification for interworking between public networks is based on Broadband Inter Carrier Interface (B-ICI) using Common Channel Signaling (CCS) extended protocol or AINI using a modified version of PNNI signaling. A number of factors influenced the technical approach to interworking among ATM based networks, and has led to the initiation and development of

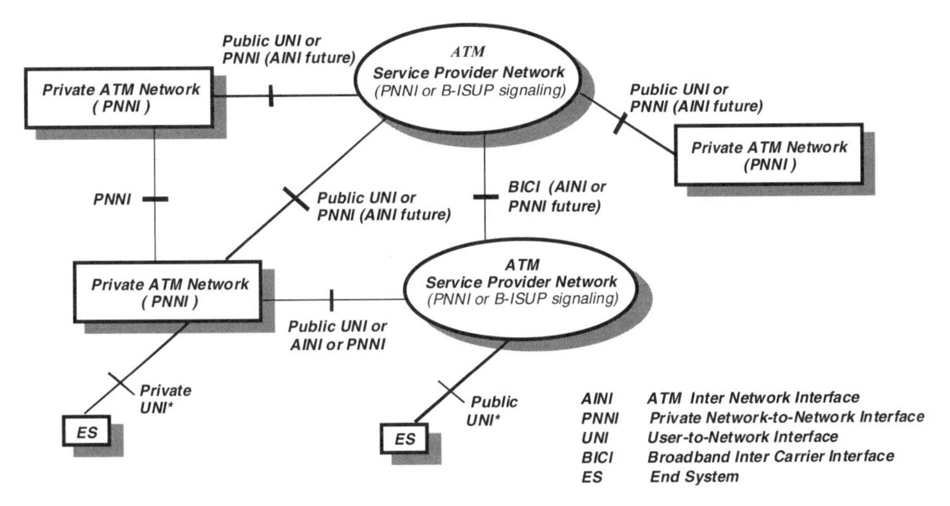

* The interface between end-user and private ATM networks and the interface between end-user and service provider ATM networks are physically the same ATM equipment. Differences are essentially in the service capabilities and features supported.

Interworking among ATM Networks
Fig. 12-2

the new interface specification call AINI. These interfaces will be discussed later in this section.

There are two types of wide area networks: private and public. While the differences are becoming fewer, some differences remain. To address these specific needs, specifications have been developed for both environments, Figure 12-2. There are several interface alternatives to base the interconnection between networks upon. This includes the User Network Interface (UNI), and the three NNI based interfaces: ATM Inter Network Interface (AINI), Broadband Inter Carrier Interface (BICI), and the Private Network Node Interface (PNNI).

The Broadband Inter-Carrier Interface (BICI) specification [30] has been developed for public carrier network interconnections. Some of the key areas addressed by the BICI are: the types of capabilities and functions needed across the interface to support the services provided to a User at the UNI; the network signaling protocol providing call and connection control information exchange from one switch to another and from one carrier network to another; and the usage parameters needed to support billing of user connections across and between networks.

As an alternative to BICI, some service providers/operators are considering the use of PNNI. The motivation is to permit dynamic routing capabilities in the public networks. However, the PNNI does not provide all the functions and capabilities that a BICI provides. In addition, there are fundamental differences in addressing/numbering. Currently the industry is split on the issue of which solution is preferred. To enable either interface specification to be used, the ATM Forum has developed specifications dealing with differences between the two interface choices which then led to the development of the ATM Inter-Networking

Interface (AINI) and guidelines on addressing and numbering. These include the ATM Forum Addressing: Reference Guide; and the ATM Forum Addressing: User Guide.

AINI describes how global end-to-end switched connections are established across any combination of interconnected private and public ATM networks. Specific architectures and reference configurations specify address formats and which signaling and routing protocols are used to support ATM end-to-end. Current discussions include what limits are to be used and what routing topology information should be shared between the ATM networks.

12.2.1 Private Network Node Interface (PNNI): The Extended Enterprise

For the extended enterprise network, the PNNI [29] specification and its precursor specification, the Interim Inter-Switch Signaling Protocol (IISP), were developed. The PNNI objective is to enable switch-to-switch, full-function, large-scale private/enterprise networks to support services already defined or adapted to ATM while facilitating new applications for corporate or enterprise networking. PNNI and associated specifications (VTOA, LANE, AMS) also provides real network consolidation benefits by providing a single wire solution. This solution is capable of, for the first time, meeting all the communication needs for any business on one platform rather than through parallel, overlay and technology-specific deployments. The savings in associated operational and maintenance costs can be significant (ranging from 25 percent to more than 50 percent over the present method of operation). In addition, PNNI offers greater performance, scalability, and flexibility in meeting the constantly changing multi-service business environment needs.

PNNI implementations are backward-compatible and interoperable with the

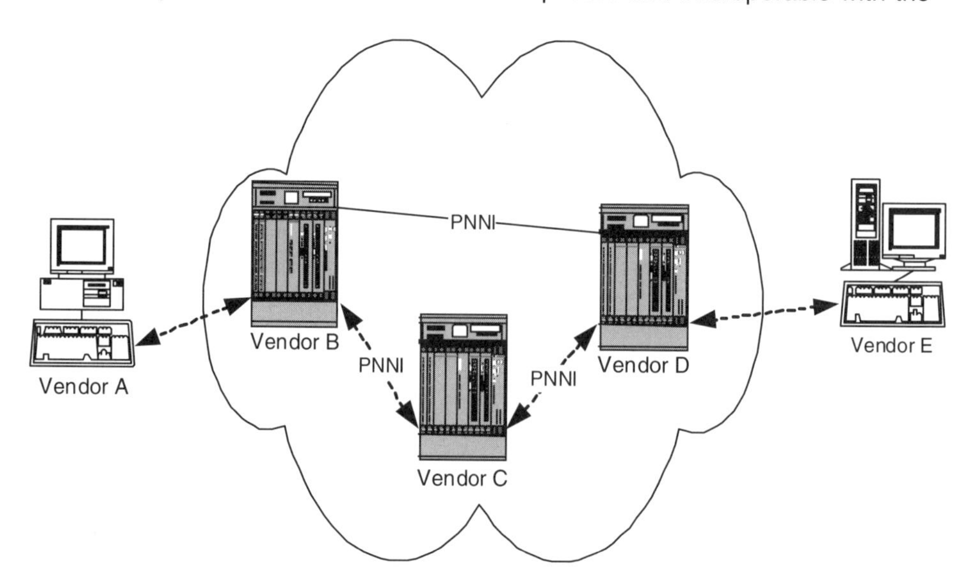

Interim Inter-switch Signaling Protocol (IISP)
Fig. 12-3

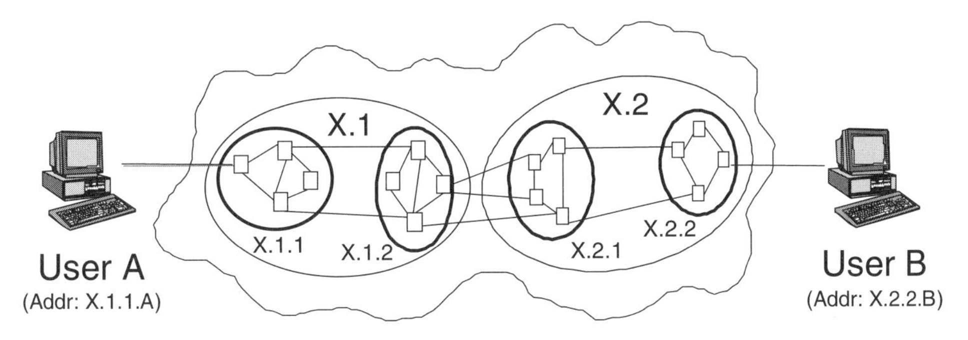

PNNI Network Hierarchy
Fig. 12-4

IISP. IISP was developed to enable small, static routed private ATM network implementations, Figure 12-3. It represented the first network SVC deployments, and ensures interoperability in small, static environments. IISP uses UNI Signaling (Q.2931 / Q.SAAL) between switches, and it requires manual configuration of static topology routing and resource tables.

The PNNI builds on the ATM UNI specification but is based on the NNI ATM cell interface. Rather than using the B-ICI specification [30] which was developed as the interface between service providers/operators and the CCS Networking Signaling Protocol, the ATM Forum decided to base its work on extending the UNI for PNNI applications. The two major technical areas addressed include extending UNI access signaling protocol for symmetrical switch-to-switch control, and developing a dynamic hierarchical state-of-the-art routing protocol. The PNNI routing protocol is able to maintain link availability information on the facilities between switches comprising the PNNI network and select the optimum route, Figure 12-4. The route calculation takes into account the type of ATM connection, bandwidth availability, and other QoS parameters. Based on the PNNI specification, much larger private/enterprise networks can be built with as many as 2000 ATM switches.

To enable the construction of these very large private networks, the PNNI routing protocol was developed to allow a hierarchy of switches logically connected in peer groups. Peer groups are interconnected through a hierarchy. The aggregation of topology state information is illustrated in Figure 12-5. This aggregation has two benefits: it is easier to configure and manage the network, and it reduces the amount of "link state" and traffic management information that has to be exchanged for effective dynamic SVC routing decisions. Within each peer group, nodes exchange PNNI Topology State Packets (PTSPs). A Peer Group Leader (PGL) is elected to represent its peer group as a single LGN at the next level hierarchy. SVCs are set up as logical links between LGNs at the same level. Public network operators are also interested in the potential application of dynamic routing and configuration to their network. However, PNNI signaling would be required as an alternative to CCS network signaling protocol. The primary limitation in this scenario is that PNNI signaling does not support the transaction-based signaling needed to gain access to intelligent

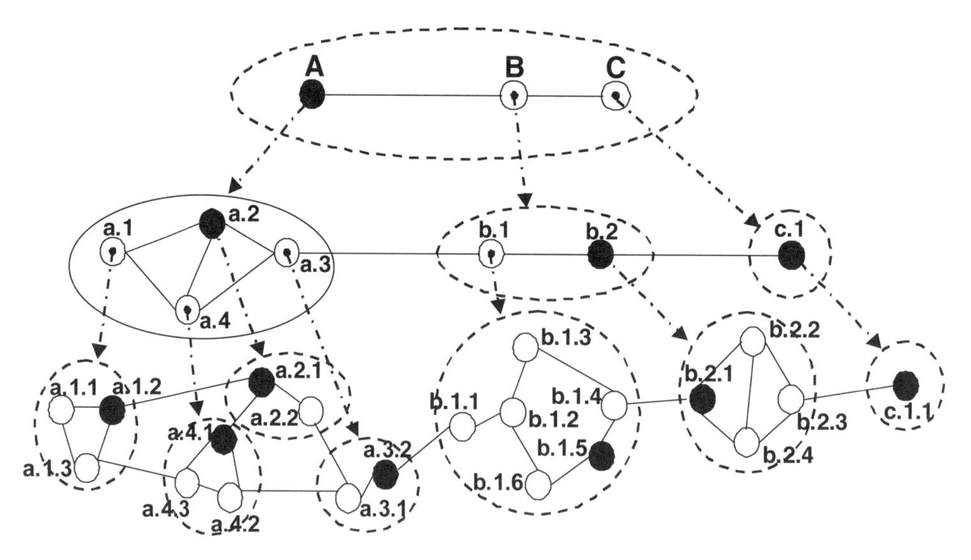

Aggregation of Topology State Information
Fig. 12-5

network capabilities and database information that CCS provides. This significantly limits added value/revenue services.

12.2.2 Public/Private Broadband ATM Interconnection

Figure 12-6 illustrates two possible views on public/private network interconnection architectures. Initially, two alternatives were available: UNI or PNNI. UNI limits the services offered to those supported by UNI. PNNI supports UNI services and allows the additional network level features developed for the PNNI to be offered. However, the PNNI also exposes one network's topology and link state information to the connecting networks. Neither a service provider/operator or corporate network manager wants to share that type of information with another network. The following summarizes the factors that motivated the development of ATM Inter Network Interface (AINI) which eliminates these problems. The goal was to:

- Support interconnection and interoperation with private PNNI-based networks without exposing internal network topology and link state information to another connecting network.
- Create new addressing/numbering techniques beyond the traditional International E.164 telephony numbering standard. This could provide a means of supporting IP applications over ATM and specific private net working needs. These are specified in UNI v3.1 and v4.0 in conjunction with PNNI.
- Development the dynamic routing capabilities in PNNI to leverage ATM's traffic management and QoS capabilities. This eases network management of statically configured networks. These equivalent capabilities and functions are not currently specified in any existing network signaling and control protocols.

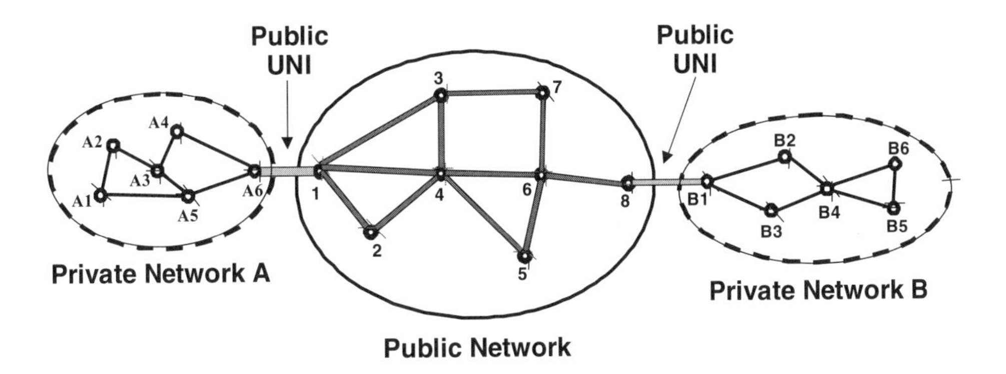

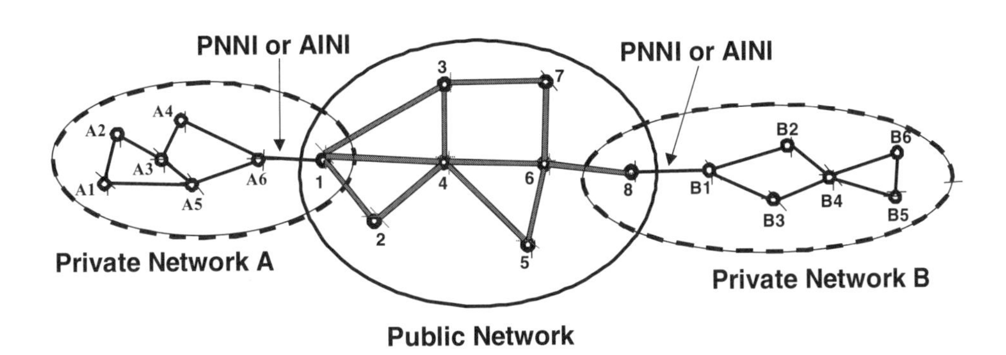

Private/Public ATM Interconnection Architecture
Fig. 12-6

- Opening up networks for competition. There was a lack of sufficient CCS protocol expertise associated with switching systems from data industry segment vendors and among a number of new service providers/operators. These vendors and service providers are considering the use of PNNI signaling as an alternative to CCS network signaling protocols.
- Allow interconnection and interoperation with connecting networks that use the BICI interface.

AINI is based on PNNI with two fundamental areas of technical difference. First, the AINI signaling protocol modifications allow interoperability between an AINI connected to either PNNI- or BICI- based interfaces. Secondly, topology state information is not passed across the interface. The AINI likely will emerge as the preferred network interconnect interface enabling interconnect of public and private networks independent of which interface the other connecting network chooses to use.

Enhanced PNNI capabilities include support of some supplementary services, extending PNNI for use in place of BICI as carrier/service provider interconnect and in support for IP to further leverage strengths of ATM. One enhancement recently completed is called PNNI Augmented Routing (PAR). PAR was developed as a method of simplifying configuration and ongoing operation of Internet level routing protocols (such as OSPF) when operated over PNNI/ATM.

PAR allows edge routers with ATM interfaces to locate other routers on the ATM network. They can directly participate in the operation of PNNI and in the embedded routing of the network (Internet or Intranet) where the router resides. With this method, the routers that have a direct ATM interface understand the full topology of the network including the routers as advertised in one protocol such as OSPF and NHRP in the future and the ATM.

12.3 NARROWBAND/BROADBAND INTERWORKING FUNCTIONS

There are a number of motivations for Broadband (BB)/Narrowband (NB) interworking, each with advantages and disadvantages. The three principle objectives are: interconnect NB switches with an ATM BB backbone, enable communications between NB and BB end users, and the provisioning of added value or supplementary services.

Interconnection of NB Switches with a BB Backbone - In this case, NB inter-switch transport facilities (DS1/E1 circuits, DS3/E3 circuits, or SONET Synchronous Transfer Mode (STM) facilities) are augmented or replaced by a BB backbone. The NB switches continue to be used to support the narrowband end users. For the public network service provider, there are a couple of benefits. First, because BB equipment has such high bandwidth, the spare BB capacity may be used in lieu of deploying new NB equipment translating into savings on the cost of NB equipment, while beginning the migration towards an all packet network. Secondly, traffic such as Internet access can be off-loaded onto the BB backbone relieving pressure on the narrowband circuit switches in the network. In this category of NB/BB interworking, a number of permutations are possible. This includes augment/replace NB facilities by BB facilities, capping NB tandem switches with BB facilities, Virtual Trunk Group capabilities modified for Public Networks, and providing a full SVC BB backbone.

Communications between BB and NB End Users - In this scenario, BB end users can communicate with NB end users. This would allow either simple point-to-point connections to occur between a BB and a NB end user or a mix of BB and NB end users on more complicated connections such as a video or audio conferencing. Possible alternatives include NB Services to BB Users via PVCs, and Basic SVC calls between BB and NB Users based partially on VTOA Desktop, VTOA Trunking, AMS, H.323 Media Over ATM, and other specifications.

Provision of Supplementary Services - It is assumed that BB end users will eventually require access to supplementary services, just as NB end users have become accustomed to supplementary services such as Custom Calling Services, Operator Services, Custom Local Area Signaling Services (CLASS), various number based services such as “800/888”, credit card calling, and Centrex business group services. If a customer is moved onto an ATM broadband switch, they will expect to still be able to make the same types of calls, with the same methods and procedures in most cases, without knowledge of the fact that they may now be served out of a packet based switch. There are several alternative approaches of providing supplementary services to Broadband users and include: Service Provision via NB/BB Interworking, BB Advanced Intelligent Network (AIN) based on Quasi-Associated Signaling, BB AIN based on Associated Signaling, NB Server Concept, and/or replicate NB Service Logic in BB Switches.

12.3.1 Interworking Function

Interworking between Broadband and Narrowband is accomplished via functionality called the Interworking Function (IWF). The IWF may be integrated into an existing Network Element (NE) or may be placed in a separate NE called the Interworking Unit (IWU).

The capabilities of the IWF are divided into two parts: user plane capabilities and control plane capabilities. (Refer to Section 4.2, B-ISDN Protocol Reference Model for discussion of these "planes".) User plane IWF capabilities are required for interworking narrowband user information to Asynchronous Transfer Mode (ATM) Permanent Virtual Connections (PVCs) and Switched Virtual Connections (SVCs). Control plane IWF capabilities are required only for interworking ATM SVCs.

At the user plane, a conversion is made between the formats of narrowband information streams (i.e., circuit mode streams based on N x 64 Kbps data rates) and broadband information streams (i.e., ATM-cell based streams at any cell rate). For 64 Kbps voice/voice-band information and N x 64 Kbps Constant Bit Rate (CBR) data (where N ranges from 1 to 24), the user plane conversion is accomplished by ATM Adaptation Layer Type 1 (AAL1) functionality.

At the control plane, there needs to be a conversion between the signaling protocols used in the Narrowband and Broadband networks. This is discussed later in this section.

In general, there are three basic types of interworking involving Narrowband (NB) and Broadband (BB) equipment, Figure 12-7. The three scenarios are referred to as Category Type 1 through Category Type 3, and the IWF units provide the following functions:

- Category Type 1: To provide interfaces for a N-ISDN network through an ATM network to another N-ISDN network
- Category Type 2: Provides interconnections between an ATM network and a N-ISDN network.
- Category Type 3: To provide support of narrowband interfaces/customer premises equipment (CPE) on an ATM network

Category Type 1 involves the use of a BB backbone network to interconnect NB switches. In this case, NB users continue to receive communications services (basic and supplementary) from existing NB switches. Consequently, it is not necessary for the ATM Broadband switches to support the full feature set that required years to develop on narrowband switches. In this scenario, it is important that the NB switches can continue to communicate without requiring NB/BB signaling interworking since it is assumed that it is not provided in this scenario. If NB/BB signaling interworking is required, then any supplementary service that relies on inter-switch signaling will not function under Category 1 interworking situations.

Category Type 2 allows NB and BB users to communicate with each other.

Category Type 1 - Interconnection of Narrowband Switches with a Broadband Backbone

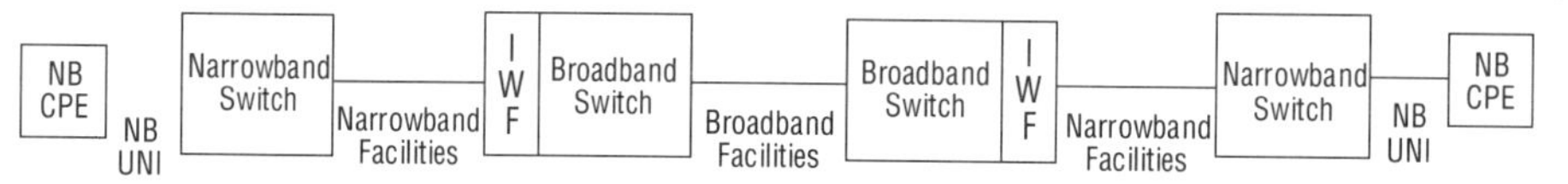

Category Type 2 - Interconnection between BB and NB Users

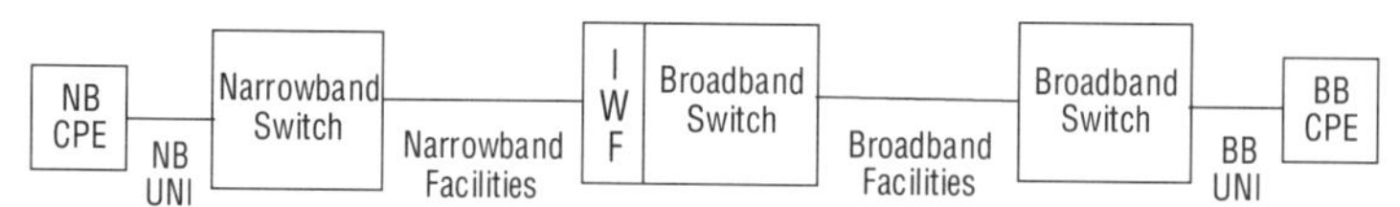

Category Type 3 - Support of NB CPE/ UNIs on BB Network

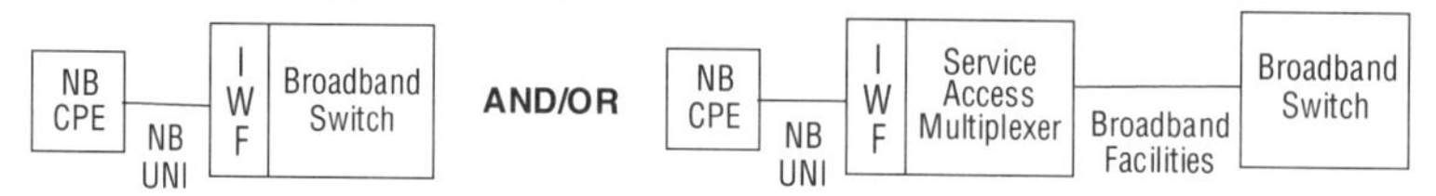

Types of Interworking
Fig. 12-7

In this case, NB users will receive telecommunications services from a NB switch and BB users from a BB switch. Category Type 2 will require NB/BB signaling interworking. Therefore, any NB added value/supplementary services that rely on inter-switch signaling will operate only if the BB signaling has comparable functionality to NB signaling.

It is important to note that BB users are likely to want comparable services as those available to the NB users. If a customer is moved onto an ATM broadband switch, they will expect to be able to make the same types of calls with the same methods and procedures. This means that the BB network needs to support NB supplementary service functionality. Currently, broadband signaling extensions to support narrowband services on a broadband interface are limited. Additional standardization work is needed. There are, however, various scenarios for signaling interworking for BB networks to support NB supplementary service functionality in advance of additional signaling standardization, and these methods will be briefly addressed later.

Category Type 3 allows a BB network to serve NB users. Category Type 3 would become important as BB Network Elements (NEs) begin to replace NB NEs. For example, if BB switches replace NB switches, there will still be the need to provide service to NB Customer Premises Equipment (CPE). One method would be to terminate NB CPE on a BB switch, while another method would be to have NEs closer to the end user (such as Service Access Multiplexers) terminate NB CPE, provide the interworking functionality, and then present a BB interface to the BB switch.

12.3.2 Signaling Interworking Scenarios

The control plane signaling interworking conversion is required between NB and BB network elements to support SVCs. When interworking occurs at the Network-Node Interface (NNI), the Narrowband signaling protocol involved in call control is CCS ISDN User Part (ISUP). For Broadband, there are three potential signaling NNI signaling protocols: CCS Broadband-ISDN User Part (B-ISUP), P-NNI signaling, and the ATM Inter Network Interface (AINI) signaling which is a subset of PNNI signaling. Therefore, there is a need to interwork between ISUP and B-ISUP, ISUP/B-ISUP and P-NNI, as well as ISUP/B-ISUP and AINI. When interworking occurs at the originating or terminating switch, the User-Network Signaling (UNI) signaling protocols from the Narrowband and Broadband network may potentially interwork with each other or with an NNI signaling protocol.

Figure 12-8 illustrates various scenarios of interworking between the different Broadband and Narrowband signaling protocol stacks. It is assumed that BB/NB interworking occurs across an inter-switch interface e.g. B-ICI, PNNI or the AINI. In addition, there may be interworking between Narrowband (NB) inter-switch or network signaling and Broadband (BB) UNI signaling. There are a number of possibilities because of different BB inter-switch signaling protocols that have been developed. This includes CCS Broadband ISDN User Part (B-ISUP) network signaling protocol, Interim Inter-Switch Protocol (IISP), Private Network Node Interface signaling, and ATM Inter Network Interface (AINI) signaling. For simplification purposes and because it is already widely deployed, it is assumed that the NB inter-switch signaling protocol will be CCS ISDN User Part (ISUP).

In Figure 12-8, it is assumed that the BB network will have both edge and

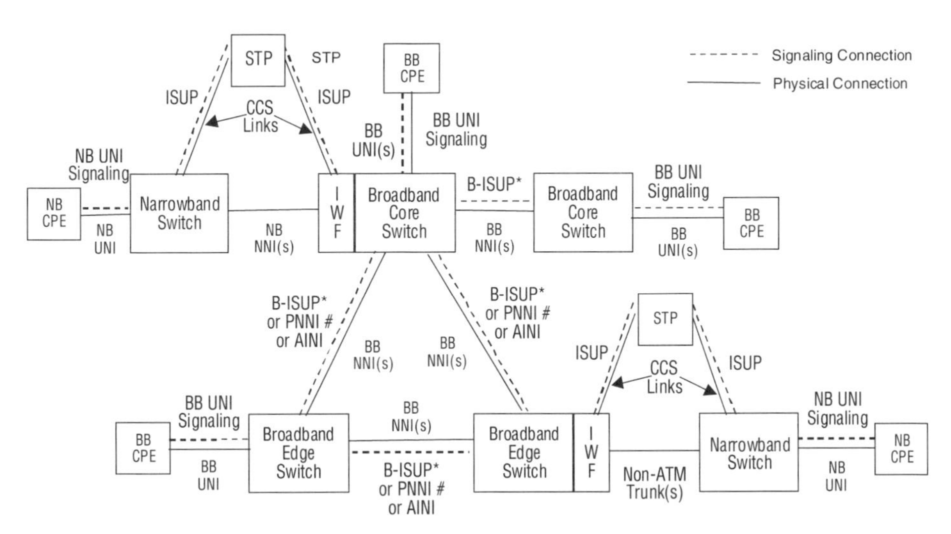

* Note - The associated mode of signaling for B-ISUP is shown to simplify the figure. The quasi-associated mode could also be used.

\# Note - If the switches are owned by different carriers/operators, then AINI would be used in place of PNNI signaling.

Different Signaling Interworking Scenarios
Fig. 12-8

core switches. Further, it is assumed that the BB network will use the associated mode of signaling initially (see Section 8), and that PNNI interface and signaling would be used between two edge switches and between a core and an edge switch. If the interface between any two switches involves different carriers or operators, than the AINI interface and signaling procedures would be used in place of PNNI.

There are four basic control plane interworking signaling scenarios between a NB network and a BB network:

- ISUP with B-ISUP
- B-ISUP with PNNI
- B-ISUP with AINI
- ISUP/B-ISUP with BB UNI

The various permutations of signaling protocol interworking are dependent on the location of the interworking function as well as the signaling protocols supported by the interconnecting networks. Additional information is available in Reference [7].

12.4 FUTURE DIRECTIONS INVOLVING ATM

A number of technical factors are significantly changing communications networking. ATM will be to the communications industry what MIPS are to the computer industry. Bandwidth price performance improvements will fuel communications growth analogous to computer growth. Digitization of information combined with ATM and computer processing power breaks down the compartmentalization which has existed between telephony, broadcast, cable, and computer networking. Bits can be transmitted over any media, readily stored, and processed anywhere. Customer devices can easily manipulate bits. How information is delivered changes. Cable Television networks are providing telephony services. The Internet is morphing from a network providing access to data to a network providing access to services. While the terminology seems minor, the shift is strategically significant with numerous new protocols being developed to ride on top the Internet Protocol (IP). These new protocols will provide capabilities and services the Internet can not provide today. All communications over the Internet are handled in packet mode. Similarly, the telephony infrastructure is morphing with a family of new access technologies for Digital Subscriber Lines and referred to as xDSL and core ATM network technology. Telephony service provider/operator networks now can offer voice, entertainment and IP data communications services.

The industry is witnessing a technological shift towards all-packet networks. One factor fueling this shift is the tremendous growth of data. In 1999, voice and data roughly consumed equal amounts of bandwidth in most North American networks. Data and voice traffic, however, are growing at different rates. Voice is growing at a rate of approximately 8 percent a year. Data, on the other hand, is growing at rates estimated between 20 percent and 1000 percent a year. Numbers vary significantly depending on the environment (business corporate network, public network) and definition of constitutes data (facsimile is data traveling over a voice network and private line/leased line service is usually

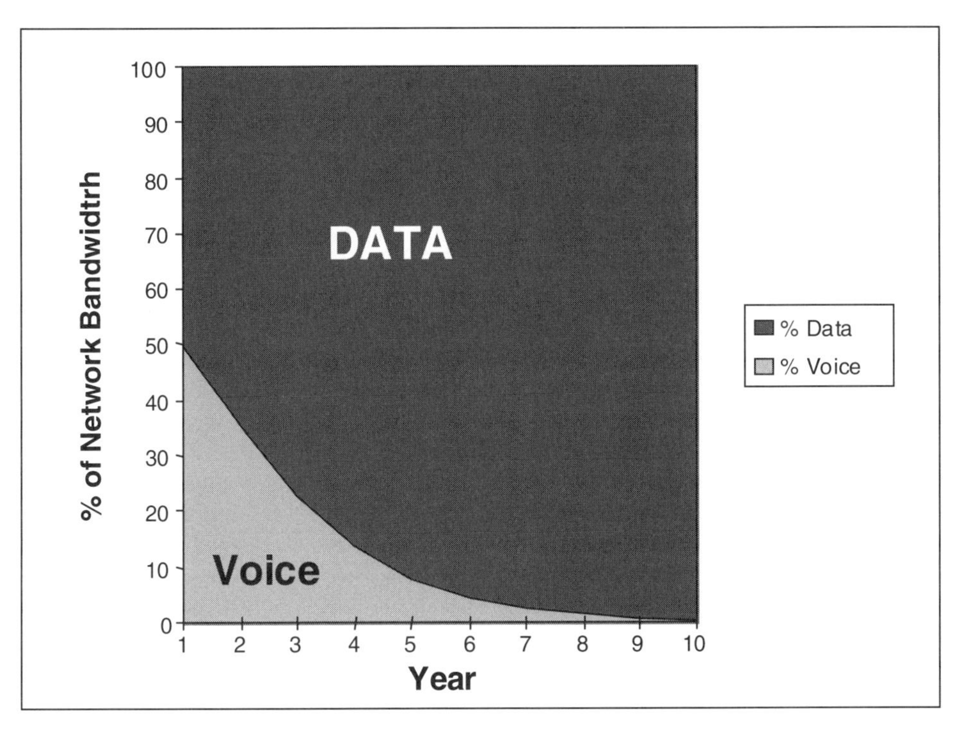

Assumption: 100% Annual Data Growth

Ratio of Voice to Data Traffic
Fig. 12-9

disguised data traffic). Using the more conservative 100 percent annual growth, Figure 12-9 illustrates that data will dominate the network in the near future.

Voice will represent a minority of network bandwidth in the future. Technology visionaries conclude that the best network planning, design and deployment strategies for the future center around supporting voice as much as possible as data. This has fueled the interest in converging voice and data networks.

Quality of service is another more important consideration that significantly influences all current industry standards, related activities, and debate regarding the best architecture and mix of technologies to support the all-packet network. While voice traffic may represent a minority of the total network traffic in the near future, voice and voice-enabled data services will account for approximately 70 percent of the revenue for a service provider/operator in the year 2004. Bandwidth itself will not pay for the cost of network upgrades to handle the tremendous growth in data.

ATM VBR packet mode provides the QoS mechanisms needed to guarantee that critical voice services are not trapped behind large packets of Web or video traffic. ATM also provides the traffic management and network management capabilities important to mission-critical applications. This is pushing the IETF to develop QoS like capabilities, Class of Service (CoS), multiple traffic classes, and signaling protocols to ride over IP protocol stack. There are some interesting intellectual challenges for the IP packet approach. IP communications is fundamentally a best effort connectionless technology. To migrate and upgrade IP so it provide QoS or "QoS-like" services has led to the development of a number of approaches including the introduction of connection-oriented packet techniques. This matches up with ATM very well.

No consensus has been reached regarding the most appropriate way of mixing technologies (routing and switching) nor on the subsequent architectural impacts of developing and delivering the network capabilities necessary to support real-time and mission-critical applications. However, most agree that the future network increasingly will be based on ATM and IP co-existence and convergence. This is a marked contrast to the IP-versus-ATM attitude of the past years.

Nearly all work underway within the ATM Forum assumes that applications will operate in the VBR packet mode. CBR will only be used for legacy interworking. This fundamental technological networking shift in approach aligns with the IP philosophy of packet-based, and now extends into the ITU-T multimedia efforts of SG 16, and supporting activities of SG 11 and SG 13.

Next Generation Networks (NGNs) refer to this packet centric paradigm shift. Combining ATM and IP enables new architectural approaches and network elements. It facilitates the migration of networks. It allows unification of middleware and provides a fundamentally different way of offering/supporting higher layer Web-driven applications. Some believe that with this architectural model, the Intelligent Network finally will be achieved.

This packet shift combined with regulatory changes opens the door for new competitive carriers/operators. New services providers/operators are looking for competitive advantages, are not encumbered with an embedded base investment that has not been fully amortized, and can leverage the lower cost technology to build a packet infrastructure.

Finally, the packet paradigm will significantly impact the existing service provider/operator base of circuit-switched networks. IP dial-up access caused problems for digital circuit switches. It changed the engineering rules for network design. The long holding time for connections for Internet use caused a significant increase in switch blocking and delayed dial tone. The impact of Internet experienced to date is the tip of the iceberg. While the shift to packet networks will not happen overnight, it likely will happen more rapidly than any previous technological shifts because of competition.

PART 2

SONET, the ATM Transport

13 CHARACTERISTICS OF SONET

Synchronous Optical NETwork (SONET) is a Physical layer, high-speed fiber optic multiplexing arrangement and set of standards. SONET standards are developed by the North American Standards Institute (ANSI) starting at the basic transport rate of 51.84 Mbps. These standards extend in multiples of that rate to at least 9.995 Gbps. The motivation to develop SONET came from the widespread and increasing implementation of fiber optic systems in both the interoffice and distribution networks.

SONET was created as a progressive hierarchy of optical signal and line rates with a robust network architecture. SONET is designed to be used on fiber optic media, which have low signal loss characteristics, yet also can be transported electrically. Also, SONET was created with robust operations features in order to improve and standardize high-speed multiplexing in networks and to enable more effective network interoperability [32].

Current ANSI standards and practices use SONET as the Physical layer for transporting ATM calls. Sections 13 to 18 of this book discuss SONET based on standards in North America.

The international version of SONET is called Synchronous Digital Hierarchy (SDH) and is based on standards from the International Telecommunications Union (ITU). The structure of SDH is similar to SONET with minimal differences. An overview of these differences is discussed in Section 19 of this book.

Why do we need a standard high-speed system such as SONET? Historically, existing transport systems have been developed in separate, or non-integrated, arrangements (such as transmission line and terminal equipment developed with no standard high-speed signaling interfaces and formats, or integrated provisioning or maintenance capabilities). As a high-speed, robust and standard system, SONET improves transmission characteristics of high-speed fiber transport systems.

- SONET provides a *transmission standard* (SONET and SDH) for high-speed transport systems at the Physical layer, creating standard optical and electrical parameters, as well as standard rates and formats.
- It improves *survivability* of the network by using a

combination of topologies, elements and frame formats to reduce the impact of errors or transmission losses caused by disruptive events in the environment.

- It uses *synchronous transmission and distribution* in the network in order to improve network transmission efficiency and survivability. Synchronous transmission provides numerous benefits in design and operation of the network, such as cost effective network equipment. For example, Add-Drop Multiplexers, which are synchronous devices in the network to efficiently add, remove or terminate traffic from high-speed networks.
- It improves performance *monitoring and alarms.*
- It improves remote *supervision and provisioning.*

To achieve these features, SONET can be established in a "physical" or hardware sense through a number of topology arrangements using various SONET network elements to configure a high-speed transport system appropriate for the application. Examples of these SONET network elements (or nodes) and topologies are SONET Add-Drop Multiplexers and fiber optic Self-Healing Rings which provide a highly reliable "highway" between SONET network elements or nodes. The combination of these topologies and elements is designed as a common signaling and transport arrangement. This arrangement includes robust management features which bring these improvements to high-speed transport.

To achieve these features in a signaling or "logical" sense, SONET establishes the physical level transport for traffic. A simple example is the transport of ATM Virtual Channels (VCs) which are carried on Virtual Paths (VPs) between customers. The ATM VP (discussed in Part 1) is a layered direct link between two ATM nodes (such as ATM Service Access Multiplexers which can be located at the customer premises). In a Virtual Channel, the ATM cells are placed into SONET transport frames and delivered across the SONET network between the two the ATM nodes.

Many VCs can be transported on a single VP, and many VPs could be carried in a physical link (SONET path). Identifiers are placed in the ATM cell headers to direct the contents of channels (Virtual Channel Identifier-VCI) and paths (Virtual Path Identifier-VPI). These identifiers are read by the two ATM network elements (or nodes, in the above example), which can act upon the cell headers and then process the ATM layer.

As other network elements are introduced into the SONET network, routing and transport efficiencies and improved survivability are available to the network provider.

This end-to-end connection links the customer to the network elements and back to the customer. Bandwidth can be assigned flexibly on a per VC basis providing a true bandwidth-on-demand capability with end-to-end VCC. Bandwidth on demand is a significant reason for implementing ATM and SONET technologies together.

Other reasons for using of ATM and SONET together include:

- A large imbedded base of SONET transport systems and Network Elements.
- A growing customer demand for ATM services.

- The cost effectiveness of a SONET system.
- Network simplification through the combination of ATM and SONET standards and systems.

The following sections address the major building blocks of SONET networks and protocols. The benefits of a SONET network and applications that support continued ATM implementation are discussed, as are several methods of transitioning from existing network systems to SONET.

13.1 SONET EVOLUTION

In the late 1980s, point-to-point SONET transport systems were introduced to take advantage of SONET efficiencies in some pieces of the emerging public optical network. In the early 1990s, SONET ring transport systems were introduced as the next stage of the evolution. By the late 1990s, initial hybrid SONET/ATM network elements (that can process both SONET and ATM layer protocols and signaling) began to appear via upgrades to existing equipment. These include:

- ATM-VP rings and other topologies.
- Add-Drop Mulitplexers (ADMs) w/ LAN interfaces.
- ADMs w/ ATM User Network Interfaces (such as DS3).
- Hybrid SONET-ATM Digital Cross-connect System (DCS).

The telecom networks of the 1980s were built on a time division hierarchy of rigidly multiplexed channels. As ATM and higher speed applications grow, these applications are being implemented in an evolving infrastructure. This infrastructure is being adapted to carry more traffic more flexibly and efficiently. It employs additional ATM and SONET elements into the overall public network, as well as various private networks.

13.2 SONET OVERVIEW

SONET provides standards for the transmission of digital signals over an optical network. SONET signals also can be transported over an electrical (non-optical) network. These signals can be both synchronous and asynchronous, so that traffic from existing networks and network components can be integrated into the optical network [32].

The standard basic SONET transmission rate is 51.84 Mbps. This basic rate is referred to as the Synchronous Transport Signal at Level 1 (STS-1) for SONET format signals, the Electrical Carrier at Level 1 (EC-1) for electrical format signals, or the Optical Carrier at Level 1 (OC-1) for optical format signals.

SONET fits into the overall transmission and signaling scheme or structure by providing a standard set of signaling protocols and physical high-speed topologies, which support higher layers of signaling and application protocols (e.g., ATM or IP). Figure 13-1 provides a block diagram of these relationships.

As illustrated, when ATM is transmitted over SONET, the ATM “rides” over or on top of SONET. Figure 13-2 shows that, at the protocol level of the Physical layer of ATM, the ATM signal is mapped (or placed) into three protocol layers of SONET (Path, Line and Section Layers). SONET is divided into these three Management Layers in order to make transport more flexible and efficient.

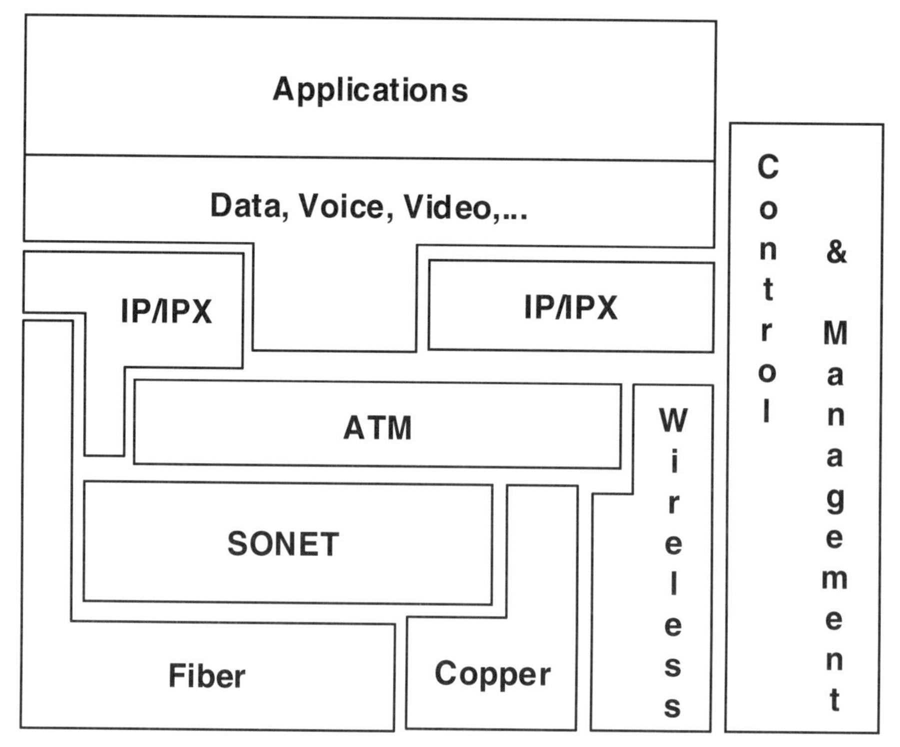

Basic Protocol Layering
Fig. 13-1

SONET Management Layers are discussed in more detail in Section 17.

As is shown in the ATM example, SONET provides the transport and topology that is a foundation for applications and higher order signaling protocols.

Using a simple, rough analogy (see Figure 13-3), SONET can be compared to portions of a train. The flatcars on the SONET train are responsible for transporting the customer payloads along with the overhead of each individual STS frame. Each flatcar represents one SONET frame payload "envelope." The

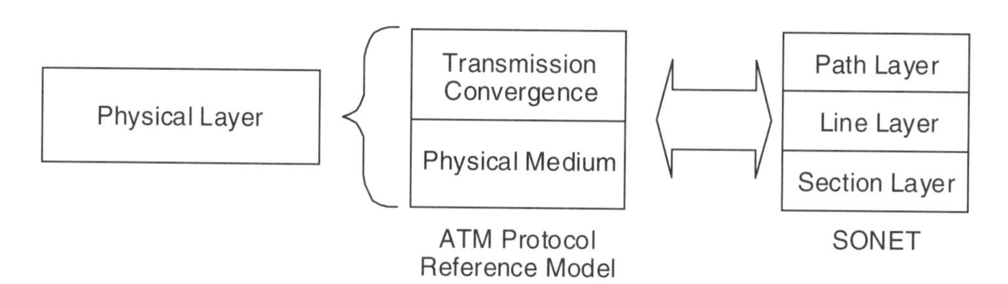

ATM-SONET Protocol Mapping
Fig. 13-2

rails upon which the flatcars ride are the fiber or coax physical medium upon which the SONET signals ride. The conductor (management layers) in the SONET train caboose (not pictured) has a view of the train to oversee what is going on. For example, the conductor can call out warnings and make observations regarding the train operation, perhaps calling for adjustments to ensure smooth and efficient transport of the cargo. ATM and IP payloads of the basic transmission hierarchy can be thought of as the more customized portion of the flatcar. For example, the customization of the flatcar frame into a box shape or tanker shape can represent these payloads.

To continue the analogy, note that the boxcar or tanker car each could handle different types of cargo. However, there is a specified but flexible volume and size structure set by the box or tank on the car. The data, voice, video and application layers are roughly analogous to the contents of the boxes and tanks (such as grain, oil, automobiles, truck trailers).

As a standardized framework for transport (the flatcar in our analogy), SONET is in an ideal position to support higher order and more flexible portions of the transmission hierarchy. Payloads for transport include ATM and IP (analogous to the boxes, tanks or payloads directly on the flatcar). Although SONET can support IP protocols directly, designers envisioned that SONET primarily would be used to support ATM protocols and applications. The ATM and SONET standards are continually coordinated to ensure that they connect seamlessly.

SONET provides a number of critical functions and features. These support ATM implementation and applications, as well as other applications requiring reliable, flexible high-speed transport. Among these features of SONET are:

- A universal optical interface, which supports the flexibility to connect to a wide variety of optical equipment.
- A mid-span meet capability, which supports the option to meet other architectures at mid-facility. A mid-span meet is the connecting of fiber cables in the middle of a span between two network operators' terminals. This arrangement promotes vendor independence, which often reduces equipment prices and supports access to best-in-class technologies. Establishing a mid-span meet capability was the first goal for the development of SONET.
- The transport of current and future services. This is accomplished through payload transparency at internal nodes, which provide the flexibility to map

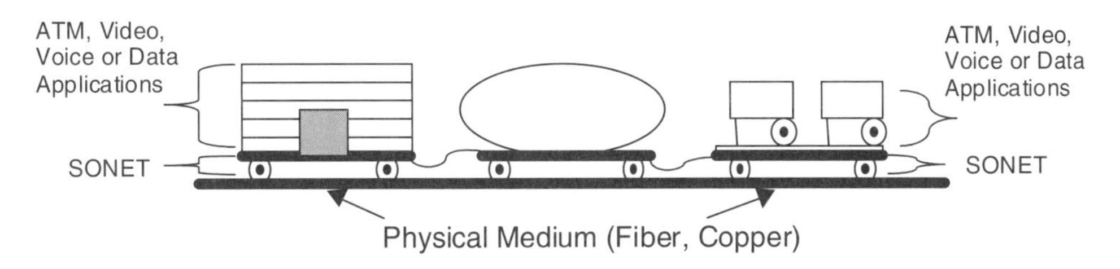

Hierarchy Analogy
Fig. 13-3

and transport different types of services (such as IP protocols and ATM).
- Extensive operations support, including fault location, error performance, monitoring and remote configuration. This also supports easy and efficient maintenance and provisioning of the system.
- Synchronous multiplexing, providing the ability to add and drop portions of the signal stream in a high-speed, low-error manner. This is accomplished through direct access to sub-rate signals and efficient add-drop multiplexing network devices. Synchronous operation (and the specific, standard formats chosen) supports manufacture of more cost-effective Add-Drop Multiplexers and fiber optic Self-Healing Rings. Multiplexers and rings are two important pieces of the overall SONET structure.
- Robust network topologies, supporting improved survivability, capacity utilization and operations. These benefits are the result of judicious placement and use of alternate (protection) paths in order to route around network problems.

To support these functions and features, SONET provides a standardized set of network elements, topologies, frame structures and management layers. In short, these SONET system parts are:
- SONET Network Elements (NEs). These elements (nodes) provide efficient interfaces at the edges of the network and to support high-speed, quality-controlled transport inside the SONET/ATM networks.
- The SONET network topology. This topology is designed to flexibly integrate the SONET NEs while supporting network survivability, which is a significant aspect in the overall effectiveness of the SONET-ATM system.
- The SONET protocol. This protocol is designed to be compatible with ATM and other protocols to provide highly efficient and flexible transport across a wide range of Network Elements.
- The SONET layers. These layers support the efficient transport, synchronization, multiplexing, maintenance, provisioning and monitoring of the SONET system. The layers of SONET are the Physical layer, Path layer, Line layer and Section layer.

The overall goal of SONET designers is to create a robust, flexible, efficient and effective network for high throughput of the types of telecommunications and information transport needed today and for the foreseeable future.

The sections that follow discuss SONET sequentially, in a building block fashion, These are:
- *SONET Network Elements (NEs).* These physical elements are necessary to transmit and receive the SONET protocol efficiently. The NEs can be thought of (using the train analogy described previously) as the physical parts of the train cars (e.g., the wheels, the steel beams of the train car frame structure). By themselves, the NEs can have a set of functions but provide much more utility when included in an overall SONET system.
- *SONET topologies.* The SONET design is made more robust by specific configurations or sequences of NEs, called topologies. Using the train analogy, this is roughly similar to a clever design of cars and train track switches which could permit, for example, adding or removing a train car to

the train while the train is in motion and placing that car on another train without delaying the car. Another analogy of the value of the SONET topology would be the rerouting of a train around a break in the track (such as moving it to an alternate track) to ensure that the train contents arrive at the final destination without significant (or any) delay.

- *SONET protocol formats.* The SONET design can efficiently carry a wide variety of payloads through the use of its protocol formats. SONET signaling formats and structure also permit a number of management functions (such as provisioning and maintenance) to occur without disrupting the overall system. A rough analogy would be a train car frame that permits a variety of different sized payloads to be carried upon it.
- *The SONET layer.* This ties together the features of all the components of SONET (NEs, topology and signaling) and describes how the overall system can be provisioned, monitored and maintained in a highly efficient manner. Using the train analogy again, the SONET layer is like the communications system built into track switches and the bar coding system on each train car. Sensors along the track read the bar codes (which constantly feed information back to the train dispatcher), so train cars can be monitored continuously and efficiently. With this information, train cars can be provisioned (constructed or added to trains to handle specified types of loads) or repair crews dispatched.
- *SONET synchronization.* Proper synchronization is critical for any digital system because transport frame or format shifts can quickly increase the transmission error rate and thus degrade the quality of service. Using the train analogy, synchronization (or sequencing, spacing and timing) is critical to avoid train crashes or derailments because the train cars are just slightly out of alignment.

SONET combines all these elements, topologies, formats and features to enable efficient transport of high-speed signals as well as the transport of lower-speed signals on an efficient high-speed network.

In order to continue the discussion of the SONET system and how it relates to previous digital arrangements, the signal rates used in the North American Digital Hierarchy (pre-SONET) and their associated bit rates are summarized in Table 13-1. This table shows the relationship between these digital terms

Signal Level	Bit Rate	Payloads
DS0	64 Kbps	1 Voice Channel or Data Circuit
DS1	1.544 Mbps	24 DS0s
DS1C	3.152 Mbps	48 DS0s
DS2	6.312 Mbps	4 DS1s
DS3	44.736 Mbps	28 DS1s

Kbps = 1000 bits per second Mbps = 1,000,000 bits per second

North American Digital Hierarchy Before SONET
Table 13-1

(developed before SONET) and their respective speeds and payload capacities. In describing SONET and its capabilities in the following sections, the signal levels and bit rates listed here will be referred to often. At the end of the discussion of SONET signaling, in Section 16, the Digital Hierarchy will be presented again (as Table 16-2). Table 16-2 will include a summary of the SONET signaling levels and payloads to demonstrate the bit rates and capacities of SONET and how they relate to the digital hierarchy shown here.

14 SONET NETWORK ELEMENTS

The basic building blocks of SONET are called Network Elements (NEs). NEs are intelligent physical nodes that transmit, receive, terminate and monitor SONET signals at various points within the network. The NE intelligence, for example, permits rapid response to network troubles by rerouting traffic around problems and/or alerting maintenance systems and personnel to troubles in enough time to avert larger network troubles.

Coupling SONET topologies and other SONET features (such as the transport frame designs and timing distributions) establishes a significant set of advantages. It creates a low-overhead, efficient transport mechanism. In addition, this transport mechanism is highly survivable and easy to monitor, maintain and provision. The combination of SONET features and its Network Elements helps ATM and higher layers of transmission maintain appropriate levels of service.

All Network Elements have three types of interfaces: transport, synchronization and operations.

Because the primary function of the NE is to transport information and data across a network, the most important interface is the transport interface. This interface is required for interaction with existing transmission equipment and facilities and is composed of both electrical and optical interfaces.

The synchronization interface receives its signal from either an external synchronization source, from the line or from the NE's own clock. The operations interface is for local maintenance technician access, alarm indications and remote operations systems.

Each NE discussed in this section incorporates all three interfaces. Specific characteristics are highlighted as the NEs are discussed below. The NEs discussed in this section are the Add-Drop Multiplexer (ADM), switch interfaces, the Digital Cross-connect System (DCS), and the Digital Loop Carrier (next generation).

■ 14.1 ADD-DROP MULTIPLEXER

The Add-Drop Multiplexer (ADM) is the most frequently deployed SONET Network Element [33] because it provides the most basic and frequently used functions. These basic functions are adding or removing a signal from the network or terminating a signal at one end of the network.

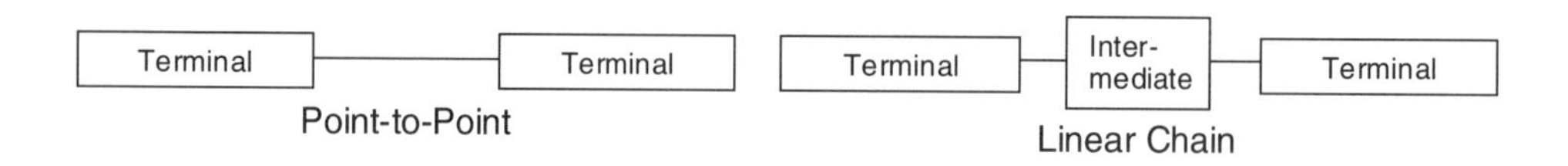

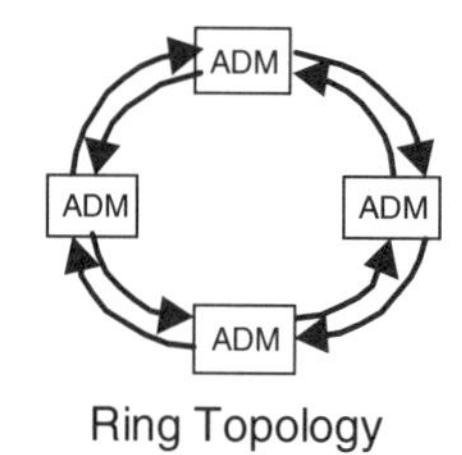

Sample Multiplexer Applications
Fig. 14-1

Terminal multiplexers were the first type of multiplexers deployed in the digital transmission plant. Terminal multiplexers have one high-speed interface and one or more low-speed interfaces. ADMs, which have more functionality than terminal multiplexers, were first used in point-to-point digital transmission systems as well as in linear chains and fiber ring architectures. See Figure 14-1[12]. ADMs have two high-speed interfaces and one or more low-speed interface(s). When SONET designers specified upgrades to the traditional ADM design, they included more robust survivability features and management interfaces to improve the usefulness of the ADM. The upgrade of the traditional ADM provides a significant part of the foundation of SONET.

ADM design for SONET needs to be simple, yet flexible, to fit a wide variety of applications and be cost effective. To this end, the ADM in a SONET network provides either one or two high-speed SONET interfaces. The ADM is arranged with one high-speed interface in the case where traffic is terminated off the network. The ADM is arranged with two high-speed interfaces to allow all or part of the traffic to pass through it.

SONET ADMs provide another basic function in the network. They aggregate low-speed traffic for high-speed transport by providing interfaces to one or more of the rates supported in the high-speed digital network. These lower-speed interfaces are known also as "tributary" interfaces.

14.1.1 Add-Drop Multiplexer Functions and Features

SONET ADMs are deigned to provide some of the basic "block and tackle" functions of the network (such as multiplexing, adding and dropping signals) as well as access to the network for supporting survivability, monitoring and maintenance functions.

ADM functions and features include:

- Multiplexing payloads from its tributary (low-speed) interfaces to the high-

[12] In Figure 14.1, terminal multiplexers are shown at the terminal ends of point-to-point applications and at the terminal ends of linear chain configurations. ADMs can be used in either the add-drop configuration (such as at the intermediate point of the linear chain or within a fiber ring) or the terminal configuration (such as the terminal point of a point-to-point circuit or linear chain).

speed interface(s). Multiplexing can be viewed as blending of one or more signal streams into a higher speed stream.

- Demultiplexing payloads from its high-speed interface(s) to the tributaries. Demultiplexing can be viewed as separating one (or more) signal streams from a higher speed signal stream.
- Deployment capability in linear chains, meshed and ring networks (see SONET Network Topologies, below).
- Cross-connection capabilities.
- Frame and Facility Maintenance capabilities.
- Operations Communications capabilities.
- Memory Administration capabilities.

14.1.2 Add-Drop Multiplexer Interfaces

The high- and low-speed interfaces of the SONET ADM are designed specifically to suit the needs of the network at the points where they are deployed.

The *high-speed interfaces* support optical signals operating at one of the standard SONET optical bit rates (i.e., OC-1 to 48 and OC-192). Thus, the ADM provides a seamless connection to the optical backbone of the network. In doing so, SONET ADM high-speed interfaces conform to the various signal format criteria for the OC-n signal, where *n* is a whole number with specified values. See standards documents: T.105 SONET Basic Description and T.117 (Digital Hierarchy) Optical Interface Specifications (SONET). ADM conformance to these standards is another building block supporting a robust SONET system design.

At high-speed interfaces, depending on the protection switching topology used by a SONET ADM, there may be additional pairs of incoming and outgoing signals. This feature supports the increased survivability of SONET compared to traditional high-speed transport arrangements in case of certain types of major network failures. For example, a SONET ADM could support (n) working signals and (1) protection signal, if the ADM provides (1:n) linear Automatic Protection Switching (APS) at a high-speed interface. When a major network failure occurs, the APS switches traffic from the working pair(s) to the protection pair(s) to route around the failure area.

The *low-speed* or tributary *interfaces* support a wide variety of electrical input signals (such as STS-1, EC-1, DS1, DS3) known as SONET or Digital Hierarchy electrical signals. The low-speed interface also may support SONET optical signals. SONET ADM low-speed interfaces conform to the various signal format criteria for the OC-m signal, where *m* is a whole number with specified values. SONET ADMs provide this support to signals at the low-speed interface so that wide range of potential signal and application types can be integrated seamlessly onto the network.

Traditionally, SONET ADMs support low-speed interfaces at full-duplex (two-way) transmission. As the popularity and maturity of SONET grows, other types of low-speed interfaces are increasingly supported. Simplex (one-way) transmission may be provided in some situations. Where simplex payloads are multiplexed through a SONET ADM, some of the functions of SONET may not be applicable (such as certain upstream signaling and error indicator transmissions).

To support the more robust nature of SONET, ADMs also provide other interfaces, which will be discussed further in Section 17 and 18. These are the:

- Synchronization Interface
- Operations Interface, which includes:
  - Local Alarms Interface
  - Craft Interface
  - Operations System Interface

The *Synchronization Interface* picks up an external timing arrangement so that the ADM is reliably and firmly synchronized with the digital NEs to which it interfaces. An example of an external timing arrangement that an ADM could use is a local central office timing source (such as Building Integrated Timing Supply–BITS). A SONET ADM may support two other timing configurations: line timing and through timing, however the through-timing configuration is not generally recommended. More information on synchronization is provided in Section 17.

The *Operations Interface* provides local and remote craft access for efficient maintenance of SONET. This interface also provides local alarm indications and interfaces to operations systems for SONET monitoring and provisioning. More information on SONET system operations is provided in Section 18.

14.1.3 Add-Drop Multiplexer: Other Features and Capabilities

SONET ADMs have cross-connection capabilities that provide flexible configurations of the ADM in various field applications. A cross-connection is the semi-permanent configuration in which two or more interfaces or entry/exit ports can be connected. The cross-connection can be done manually (by a maintenance technician), or the ADM can automatically assign them when configured to terminate traffic. Also, the cross-connection can be done remotely, if the ADM is so equipped.

For example, an ADM can use a cross-connection to connect a high-speed port to a low-speed port to allow a portion of the traffic to exit the network at that ADM. The ADM reads the address, recognizes the traffic to exit and removes it from the high-speed transmission stream. Then, via the cross-connection in the ADM, it delivers the traffic to the low-speed exit port on the ADM for delivery to the tributary network.

Currently, six types of cross-connections have been defined for use by SONET ADMs [33]. The first three types are most commonly used. The last three types are more specialized and would appear less frequently in the network. See Figure 14-2. The six types are:

- Add-Drop
- Through
- Drop and Continue
- Hairpin
- Multi-Drop
- Multi-Drop and Continue

In an *Add-Drop* cross-connection (sometimes referred to as a "drop" cross-connection), a received path at one of the SONET ADM high-speed interfaces is demultiplexed and dropped to one of its low-speed interfaces, where it may or may not be terminated. The reverse operation is performed on the incoming

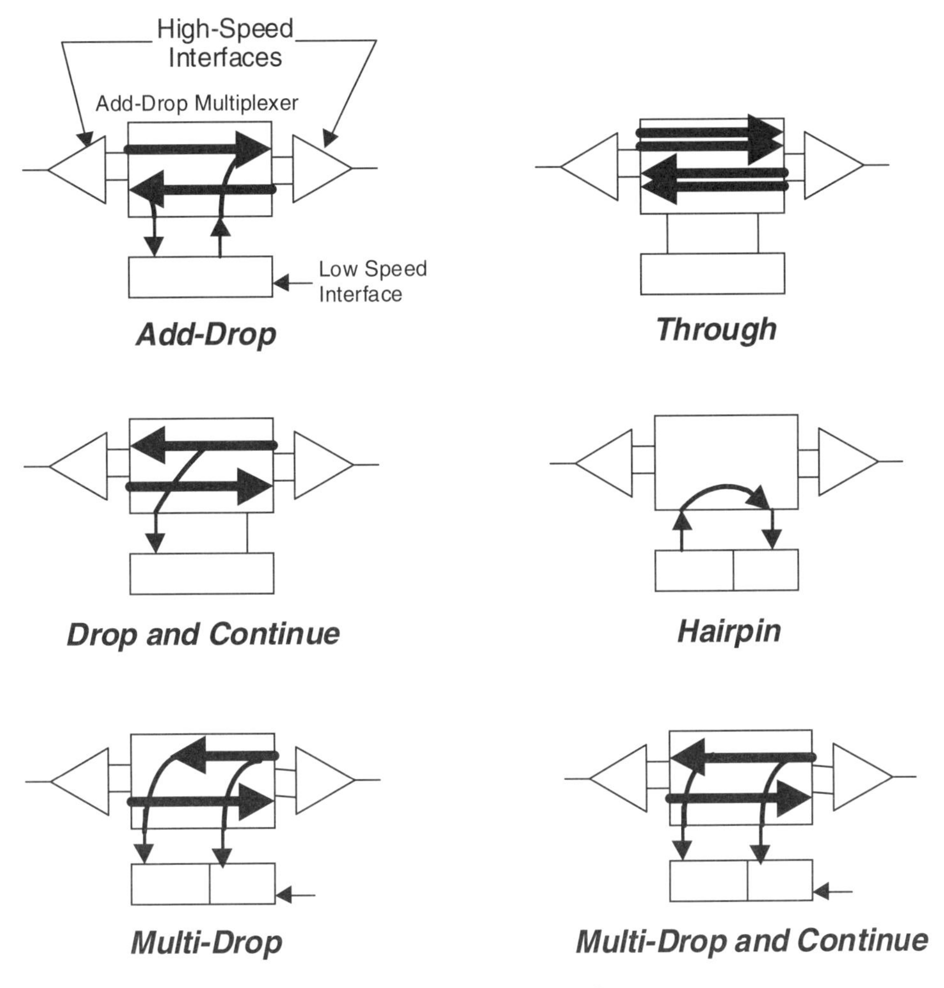

Add-Drop Multiplexer Cross-Connects
Fig. 14-2

signal at the tributary interface and is called an "add" configuration. The Add-Drop cross-connection supports one of the most common applications of the ADM, which is to add or remove traffic from SONET.

In a *Through* cross-connection, a non-terminating receive path at one of high-speed interfaces is passed through to the outgoing signal at its other high-speed interface. This cross-connection can be used, for example, when a number of connections are presented to the ADM and a portion of them need to pass through this particular Network Element transparently in order to be terminated elsewhere on the network.

A *Drop and Continue* cross-connection establishes the connection for a received path, at one of the high-speed interfaces, to be dropped simultaneously to a low-speed interface (where it either may or may not be terminated) and continue on to the other high-speed interface. Typically, this cross-connection is used in a broadcast application. Another application for this cross-connection is to interconnect fiber rings.

Hairpin cross-connections establish the connection for a received signal or path at one of the low-speed interfaces to be connected to an outgoing signal or path at another of its low-speed interfaces. Each low-speed interface is usually a two-way cross-connection, with at least one of the interfaces being a SONET interface. The Hairpin cross-connection has only recently been defined in the industry and will continue to be established in new applications in the future.

A *Multi-Drop* cross-connection allows the signals on the receive path from one of the high-speed interfaces of the ADM to be dropped and then bridged to two or more of its low-speed interfaces. This cross-connection is applicable only on a broadcast, or one-way, basis.

A *Multi-Drop and Continue* cross-connection is a variation of the Multi-Drop cross-connection. In this case, the received path at one of the SONET ADM high-speed interfaces is also passed through to the other high-speed interface. This type of cross-connection typically is used for one-way or broadcast applications.

14.1.4 Add-Drop Multiplexer Configurations

SONET ADMs have two configurations (or modes): add-drop and terminal. Since the SONET ADM is arranged with these two configurations, and these configurations are designed to accept a broad range of inputs, the ADM has become one of the major "work-horse" Network Elements of the SONET system.

In some cases, a single ADM may be capable of operating in either the terminal or add-drop configuration. This capability contributes to the flexibility of the ADM and, thus, increases its frequency of implementation in the network.

Figure 14-3 shows the basic functional architecture of the ADM.

The *add-drop configuration* of the ADM is used in many different network topologies. It has a unique Network Element architecture, which eliminates the need for back-to-back multiplexing. Back-to-back multiplexing occurs, for example, when a data stream (or channel) needs to be separated from a higher-rate data stream for redirection. Then, the single data channel is extracted, and the remaining data stream is multiplexed again and sent on its way. Without ADMs, the entire data stream has to be demultiplexed.

ADMs in the add-drop configuration, for example, can insert or drop individual DS1 or DS3 (or other types of) channels into the OC-N stream. The SONET add-drop configuration ADM provides two high-speed interfaces. These interfaces are commonly referred to as "east" and "west", or "1" and "2". Similar to the terminal configuration, add-drop configuration may drop payloads between each high-speed interface to the low-speed interfaces. In addition some payloads contained in the incoming signal may be passed through to the other high-speed interface.

The add-drop configuration of the ADM also provides management and survivability functions (such as path management, dynamic bandwidth allocation, channel protection) for the overall SONET system. This configuration contributes to efficient processing of traffic flow and may result in a design requiring fewer total Network Elements.

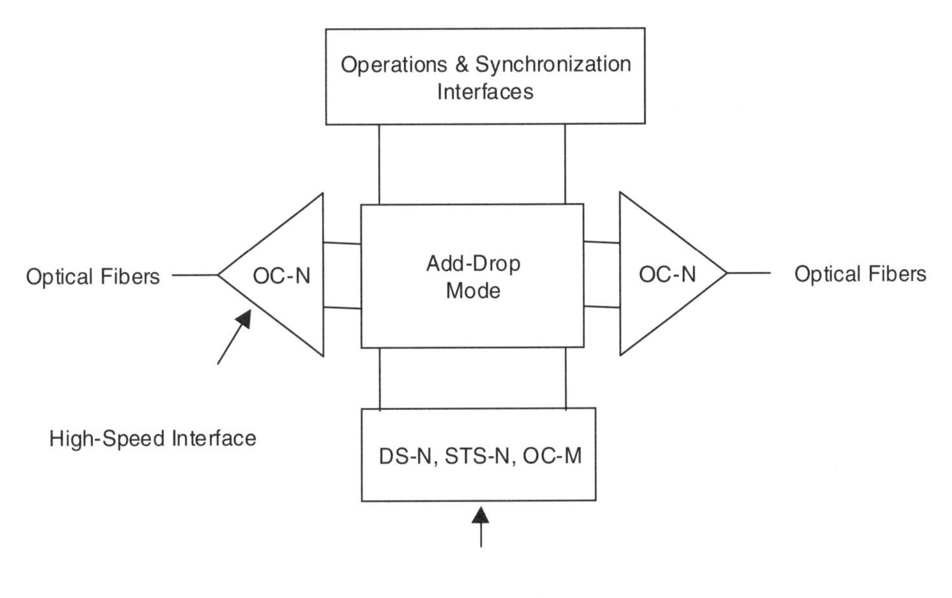

ADM Functional Architecture
Fig. 14-3

The *terminal configuration* of the ADM is a special case of the add-drop configuration. In this configuration, the SONET ADM provides a single high-speed interface and all of the payloads from the incoming high-speed signal(s) are dropped to low-speed interfaces. Also, all outgoing signals are multiplexed to the high-speed interface from the low-speed interface(s). The terminal ADM adds or drops all traffic presented to it. With these functions, the terminal ADM can perform the function of service aggregation from multiple broadband service access interfaces (such as DS1/DS3, video, LAN, ATM CRS) to a single high-speed interface to the SONET network.

14.1.5 Add-Drop Multiplexer - Applications

Because the ADM is the most frequently deployed SONET NE, a number of its applications will be discussed in some detail.

14.1.5.1 ADM Applications - Background

When SONET deployment began, the network environment contained asynchronous multiplexing dominated by the DSn hierarchy (Digital Hierarchy shown in Table 13-1). SONET NEs had to coexist in the network with existing equipment, such as asynchronous multiplexers. Therefore, the initial function of most SONET ADMs was to map and multiplex DSn signals into OC-N signals, primarily via asynchronous DSn mappings. This is still a major application for SONET ADMs.

However, as SONET becomes more widespread, there is an increasing need for ADMs to support the growing number of types of SONET low-speed interfaces.

These include low-speed interfaces for the distribution, routing and grooming of SONET-mapped payloads through large SONET networks. In order to support the increasing types of low-speed interfaces, designers arranged the SONET ADMs so that they can operate at different OC-N high-speed interface rates connecting to a variety of low-speed interfaces. For example, Terminal ADMs at either end of an overall end-to-end connection may have different low-speed interfaces for the different customer needs.

In addition, SONET ADMs support payloads containing ATM traffic which provide the foundation to support a wide variety of ATM applications discussed in Part 1.

SONET ADMs are used in any of these four topologies:

- Point-to-point (single OC-N rate)
- Linear chain
- Hub
- Ring

The *point-to-point* topology is an arrangement between two points for traffic transport between two corporate locations, as an example. The *linear chain* is a topology where Network Elements (such as ADMs) are linked together sequentially between several points. In a linear topology, the far ends of the network do not connect in a circle or ring. A *hub* topology is dominated by a center or hub to which each end point is directly connected. A *ring* topology is similar the linear chain, except the end points are connected to form a circle. Figure 14-4 provides a basic diagram of these four ADM topologies.

14.1.5.2 ADM Application Examples

As SONET expands throughout the telecommunications and data networks, more applications are being implemented on an ongoing basis. In addition, with

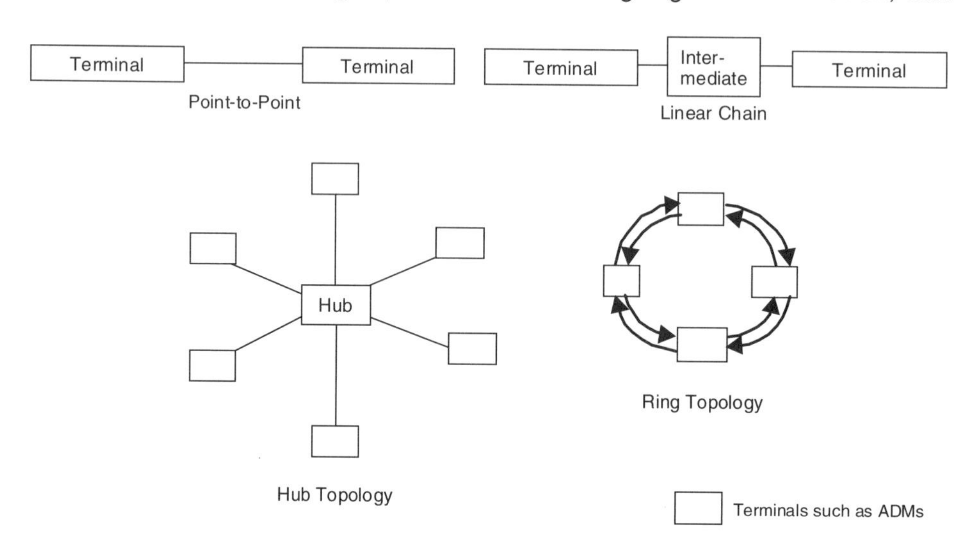

Topology Applications
Fig. 14-4

the implementation of SONET in non-traditional telecommunications and private networks, more applications for all of the Network Elements are being discovered.

Four specific ADM application scenarios are common in the network today. They are the:

- Interoffice Network Application
- Loop Application
- Hybrid Traffic Application
- Service Access Multiplexer Application

Interoffice Network Application

SONET ADMs are used in the connections between central offices in any of the four topologies noted previously. The most popular is the ring topology. In a simple interoffice network application with the ring topology, the ADMs are deployed on the ring. Central offices are connected to each of the ADMs. Traffic from a central office is multiplexed onto and off the ring via the ADMs.

Other interoffice applications include the linear chain architecture. In this topology, central offices are connected using a terminal ADM on each end and one or more add-drop ADM(s) in between. Typically, low-speed interfaces from the ADMs in this application are DS3, or OC-M.

Loop Application

SONET ADMs also can be used in access loop applications. The motivation for using high-speed connections in the access loop is based on economics. The cost of installing a typical telephony two-wire loop from the end office to the subscriber can exceed $1500. The bandwidth, which can be carried on an unmodified access loop (where high-speed digital or other electronic circuitry is not employed on a copper pair of wires), is typically only enough to support one voice circuit (such as a DS0). However, when higher-speed digital technology is employed, many multiples of the capacity of an unmodified loop can be achieved. For example, a DS1 circuit (using two pairs of wires) has the capacity to transport 24 voice circuits. This higher speed saves considerable loop installation and maintenance costs and permits higher bandwidth applications, such as high-speed data and video, to be transported at a reasonable cost.

For example, SONET ADMs in the terminal configuration could be deployed by placing one ADM at the central office (near-end, in this example). Another ADM is placed close to the end of the loop plant (far-end, in this example) and connected to a Remote Digital Terminal (RDT). See Figure 14-5. OC-N or DS1 signals are transmitted from the Central office to the near-end Terminal configuration ADM from the Central Office Terminal (COT) for entry onto the SONET portion of this network. At the far-end ADM, another Terminal configuration ADM sends outgoing OC-M or DS1 signals to the RDT for final termination in the network.

In some loop applications, SONET ADMs can provide transport for SONET-based Digital Loop Carrier (DLC) systems. See Figure 14-5. Two types of Digital Loop Carrier systems are typical in a SONET system. These are the Universal DLC (UDLC) system and the Integrated DLC (IDLC) system.

The UDLC consists of RDTs and Central Office Terminals (COTs). SONET ADMs generally will transport these RDT-COT signals transparently (using asynchronous DS1 mapping, for example).

In an IDLC system, the RDT communicates with the Integrated Digital Terminal (IDT) of the switch (Local Digital Switch - LDS). The IDT incorporates the COT features and, thus, a separate COT is not needed. The SONET-based RDT-IDT traffic is transported transparently using byte-synchronous DS1 mapping.

Hybrid Traffic Transport Application

In some cases, a SONET ADM can transport hybrid traffic (such as ATM and

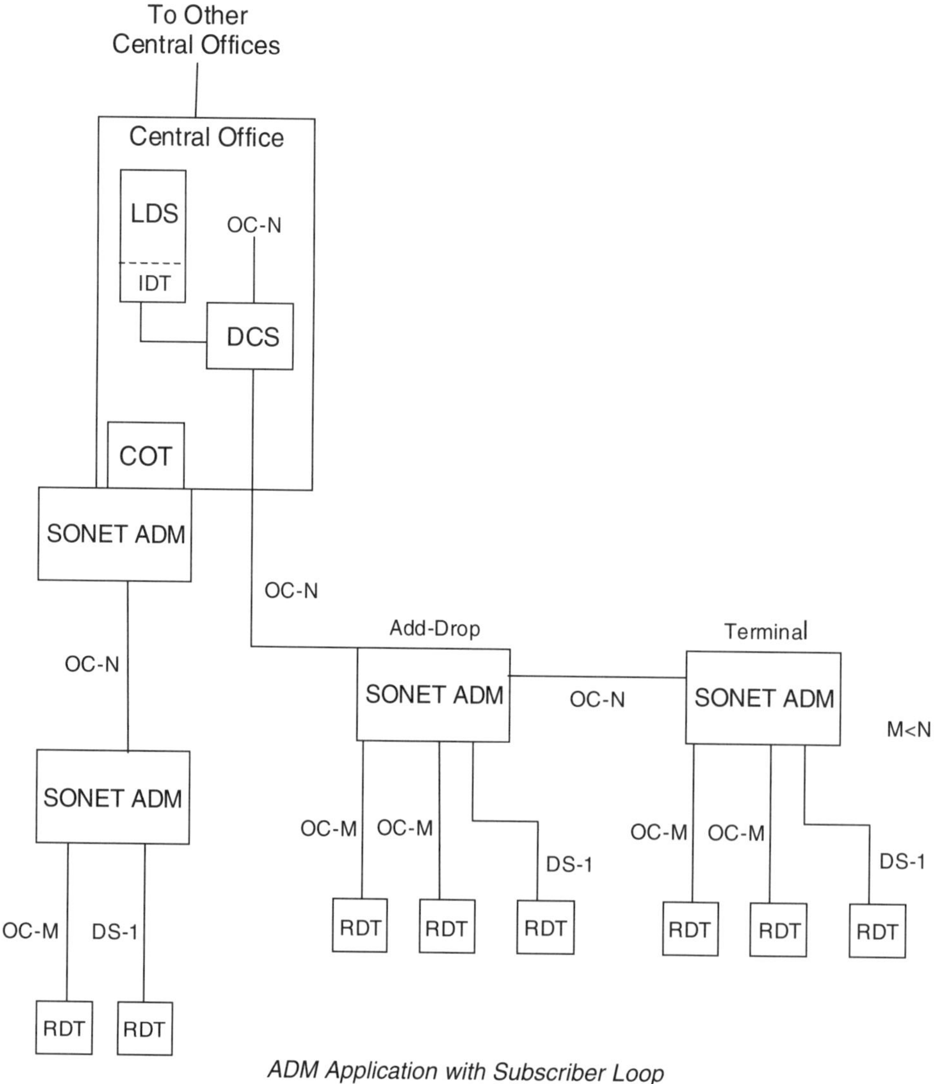

ADM Application with Subscriber Loop
Fig. 14-5

traditional Synchronous Transfer Mode, or STM) over the same SONET pipe. Transporting hybrid traffic can be an element of a plan to transition from an existing pre-ATM network to a fully functional ATM network. Techniques used to transport hybrid traffic, include:

- Transporting both STM and ATM traffic over the same pipe, but in different paths. See Figure 14-6 for an example of this application [40].
- Converting the STM traffic that does not contain ATM cells to ATM cells via ATM Circuit Emulation Service (CES) and then carrying it and other ATM traffic in a single path.

Service Access Multiplexer Application

The ATM Service Access Multiplexer (SAM) can be integrated into a SONET ADM to perform a network edge function. The SAM can aggregate traffic from multiple broadband service access interfaces and/or customers from the edge of an ATM network. This provides a cost-effective broadband service access

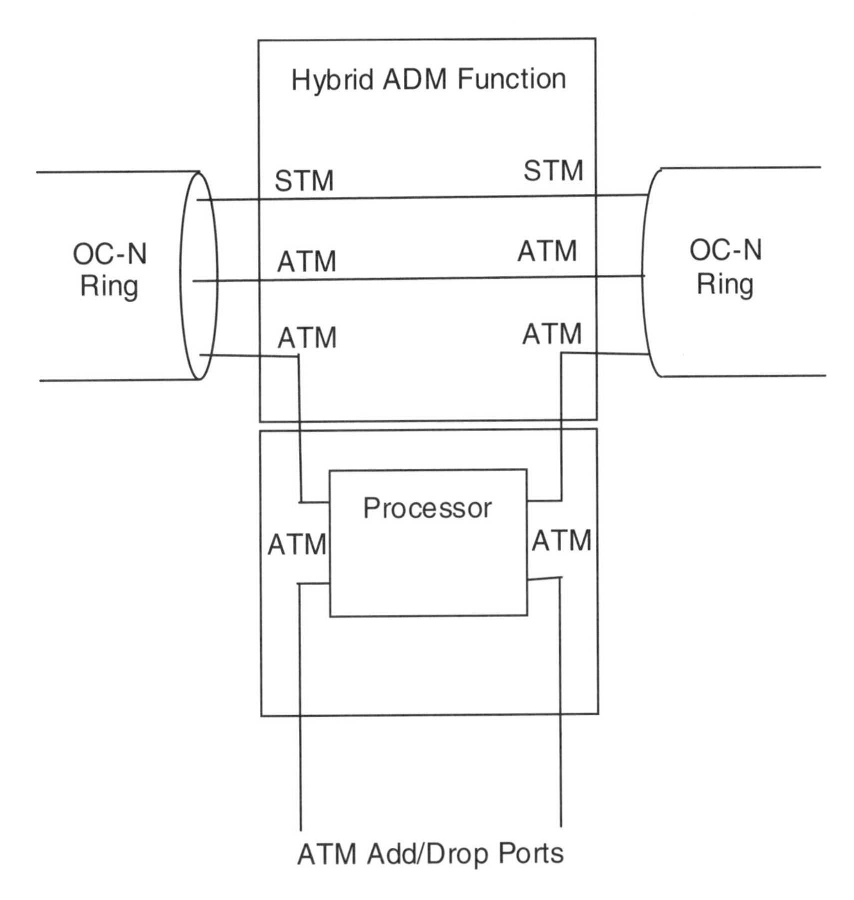

STM and ATM - Hybrid Traffic Application
Fig. 14-6

platform. The SAM function is the broadband equivalent of the remote digital terminal in a conventional telephony network.

Hybrid arrangements such as these can assist operators to more efficiently transition to an ATM and SONET environment while making best use of an embedded base of transmission networking equipment.

14.2 SWITCH INTERFACES

SONET provides for voice and ATM switch interfaces. This makes SONET a flexible transmission technology for telecommunications architectures of today and tomorrow. This section will discuss these switch interfaces and examples of their applications.

14.2.1 Voice Switch Interfaces

SONET can provide an efficient, high-speed connection between voice switches (trunk side), or between a voice switch and another network topology element, such as a fiber optic ring or a Remote Digital Terminal (line side). Traditional voice switches are connection-based, meaning that the connection is established for the duration of the call. However, before entering SONET, voice (or voice circuit switched data) traffic is converted to digital signals. Upon entering SONET this digital traffic is placed in SONET frames for transport.

Before discussing SONET and voice switches further, it is useful to understand the architecture of the typical local telephony voice network. In the typical local telephony network there are two types of central offices or switches: end office and tandem. See Figure 14-7.

An *end office* voice switch has connections to subscribers, these are called lines. Connections which link to other end offices or tandem switches are called trunks.

Tandem voice switches connect end offices to other end offices or to other tandem switches using trunks.

The *line side* of an end office voice switch typically interfaces to individual subscriber lines through analog or digital (DS0 or DS1) connections, often via metallic cable pairs. When connecting to the line side of a voice digital switch, the SONET network interfaces to the line side digital termination module of the switch (via a Central Office Terminal, for example). Then SONET interfaces at the far-end of the line via a Remote Digital Terminal.

The *trunk side* of an end office voice switch contains trunks that transport many subscriber calls or connections. A voice trunk is usually a common transport, connecting traffic for whichever subscriber needs traffic capacity to complete a specific call. In the traditional circuit-switched telephony network, the trunk is fully dedicated to a call for the duration of the call. Trunk circuits interconnect voice switches in the network. A trunk circuit could be carried over a variety of facilities and transport topologies (such as hub or ring). To connect to the trunk side of a voice digital switch, the SONET interface connects traffic (converted from analog to digital, as necessary, and then to SONET formats) between switches in the network or across networks.

14.2.2 ATM Switch Interfaces

The ATM switch primarily employs SONET interfaces to provide its transport

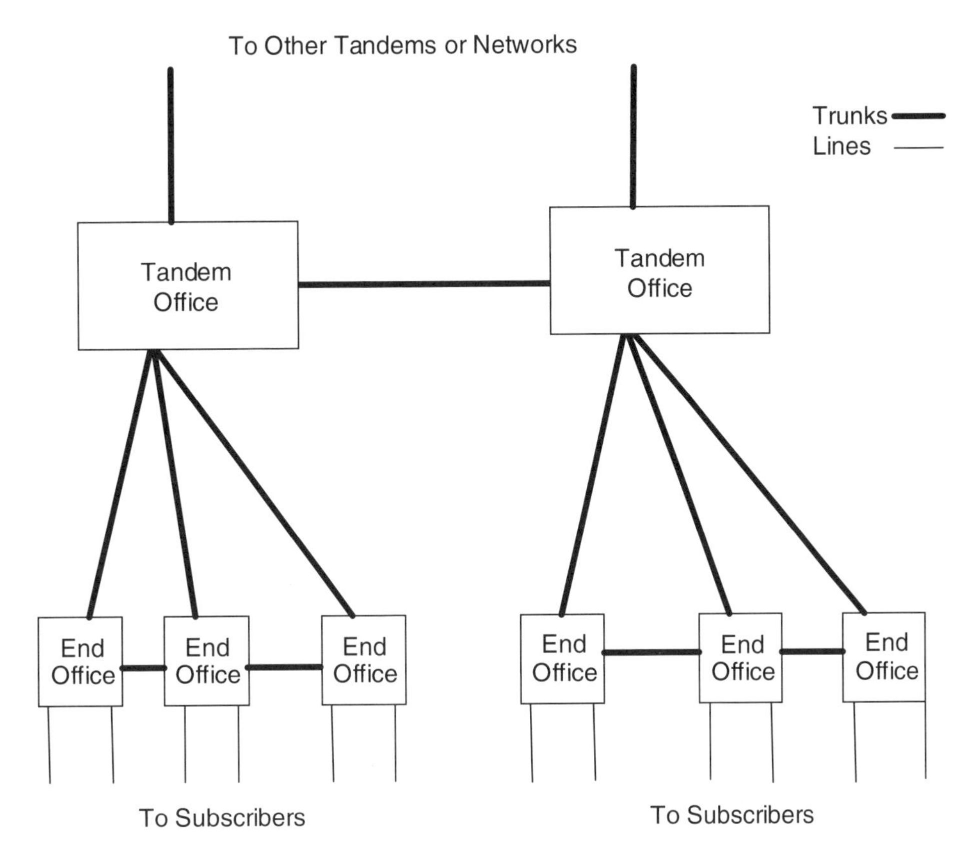

Traditional Local Telephony Voice Switch Architecture
Fig. 14-7

using a variety of standard rates. SONET supports high levels of throughput for broadband services on the line side of the ATM switch interface. SONET throughput rates for the ATM switch line side are: Synchronous Transport Signals Level 1 (STS-1), STS-3c and STS-12c (51.849, 155.520 and 622.080 Mbps respectively), where “c” denotes concatenated, or a combined frame format. The trunk side throughput rates are STS-3c, STS-12c and STS-48c (155.520, 622.080 and 2488.32 Mbps respectively), which are well suited for the ATM switch. SONET provides other benefits such as performance monitoring and alarms. The ATM switch can access these features directly through its SONET interface.

There are two types of ATM switches, the Edge Switch (ES) and the Broadband Switching System (BSS). Both types of ATM switches have interfaces to SONET.

The largest type of ATM switch is the BSS. The BSS is used as an ATM hub switch. Hub switching, (see the hub concept in Figure 14-4), is most suitable for connections between different and distant customer fiber rings.

In addition to ATM traffic, a BSS can also handle traffic (and has interfaces) for other non-ATM services, such as Switched Multimegabit Data Service (SMDS), DS1/DS3 Circuit Emulation Service and PVC Frame Relay Service. A typical BSS installation would include ATM traffic and some or all of these other services.

The ES is used at the edge of an ATM network. The ES has less capacity than a BSS. It is most suitable for connections within a customer's fiber ring or between connected customer fiber rings, however, it can handle longer-distance connections as long as they are not dominating. The ES can be used along with the Service Access Multiplexer (SAM) to provide a low-cost broadband service access platform.

The ATM part of a carrier's network could be based on a few large hub switches or a larger number of small edge switches. Depending on circumstances, one or the other may be more economical. Figure 14-8 and 14-9 show the hub and edge switching architectures, respectively.

These figures show the hub, edge switch interfaces and SAM in different topology examples. These figures also show a simple example of how ring topologies can be used with these interfaces.

Large networks that employ a number of rings (and thus, would use Hub and Edge switches) can be organized in a hierarchical manner. The basic architecture of the broadband network was introduced in Sections 10-12. Access (or customer) rings represent the lowest level in the hierarchy. Interoffice (or local,)

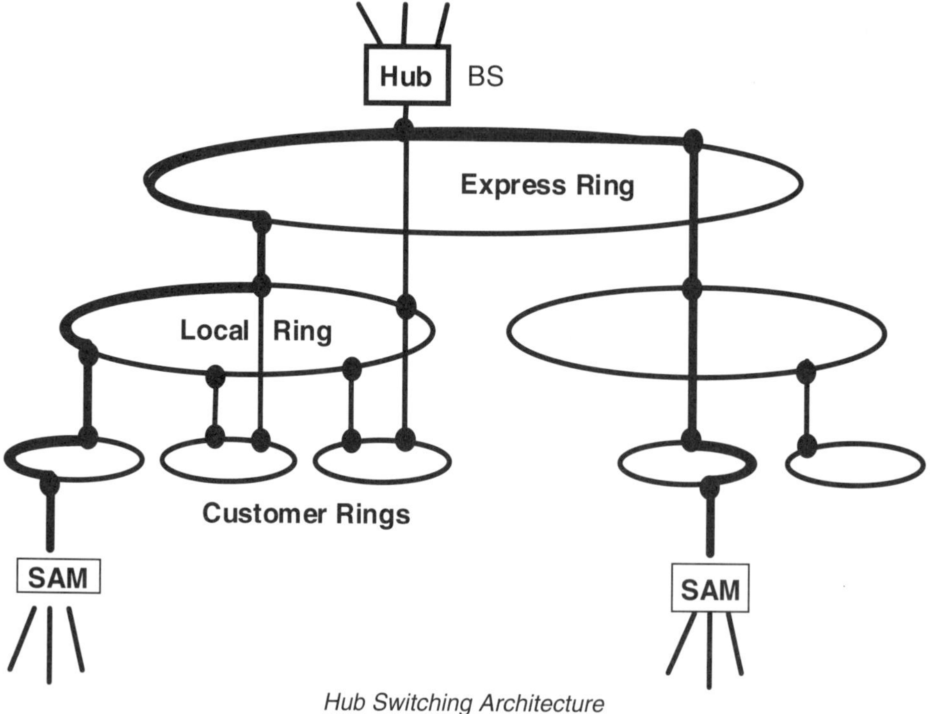

Hub Switching Architecture
Fig. 14-8

and backbone (or express) rings represent higher levels, as shown in Figures 14-8 and 14-9. The SONET express ring transports traffic between local rings and, at times, to and from customer rings. Local rings transport traffic between customer rings and from customer rings to the express rings. The customer rings transport traffic from customer Network Elements to the local and express rings. Figures 14-8 and 14-9 present ring topology connections using the hub and edge switches to Service Access Multiplexers as a basic example of the interfaces and relationships between these NEs. SONET ring topology will be discussed further in Section 15, below.

The SONET network relies on ATM switches (ES and BSS) as well as the SAMs and the NEs in the network to carry out ATM processing functions. [31]

Services supported by the typical BSS can include:

- Permanent Virtual Connection (PVC) Cell Relay Services (CRS)
- Switched Virtual Connection (SVC) CRS
- Switched Multi-Megabit Data Services (SMDS)
- LAN interconnection
- PVC Frame Relay Service (FRS)
- DS1/DS3 Circuit Emulation Service (CES)
- Interworking between broadband and narrowband networks for some limited capabilities, such as voice over ATM or 64 Kbps data over ATM

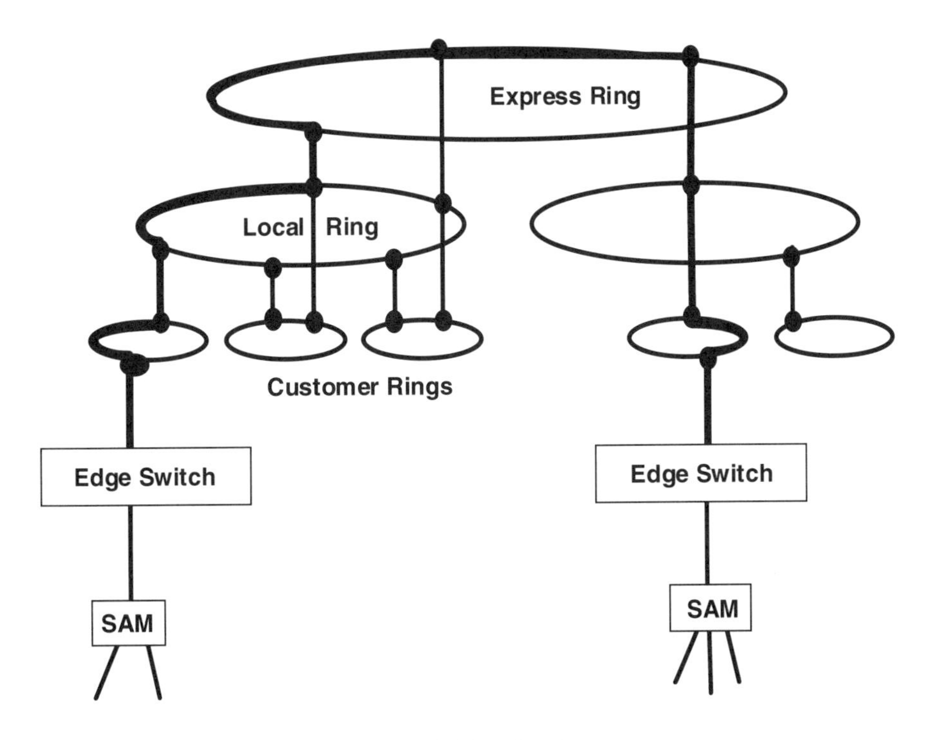

Edge Switching Architecture
Fig. 14-9

ATM switches also support traffic usage measurement for network management (such as traffic control and congestion management functions) and input to the billing function. This supports the overall Quality of Service (QoS) for the combined ATM/SONET network.

14.3 DIGITAL CROSS-CONNECT SYSTEMS

A *Digital Cross-connect System* (DCS) is a major SONET Network Element. It provides a means to connect digital paths from numerous interfaces in a semi-permanent condition, which can be changed using software commands. The connection paths also can be dynamically allocated in the SONET DCS. The SONET DCS is designed to be highly effective in applications with large volumes of traffic and intensive connectivity.

In addition, the SONET DCS can provide functions such as add-drop, grooming, broadcast, facility rolling, facility performance monitoring and test access. These DCS functions add considerable flexibility to service provisioning and maintenance of a SONET system implementation.

14.3.1 Digital Cross-connect System Functions

14.3.1.1 Digital Cross-connect System Basic Function

A major function of a DCS is to interconnect SONET rings into a mesh

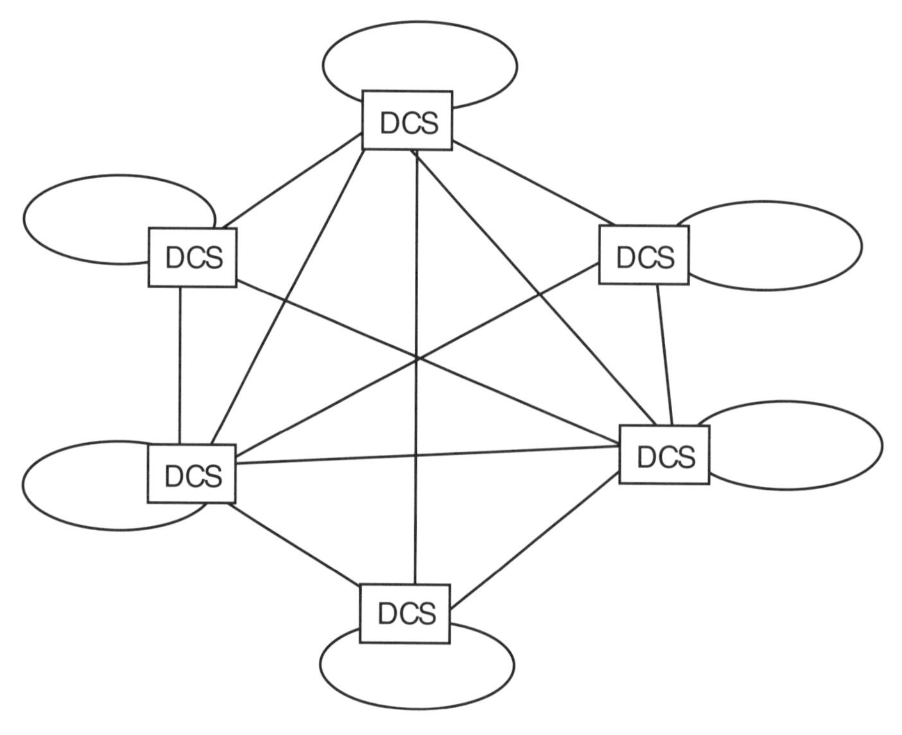

Mesh Topology
Fig. 14-10

network topology. See Figure 14-10 for a conceptual example of a mesh network. Interconnection is implemented by rearranging the SONET paths in a DCS so that they can be dynamically routed to the appropriate destinations. See Figure 14-11 for a functional architecture of a DCS.

14.3.1.2 Additional Digital Cross-connect System Functions

In addition to performing intensive high-volume traffic connectivity, the DCS provides many traffic routing, management and maintenance functions. This flexibility makes the DCS highly desirable in high-traffic volume networks where more complex traffic routing is required. This section describes the major functions of the DCS beyond its basic cross-connect capability.

The *add-drop function* supported by a DCS is the same capability provided in the Add-Drop Multiplexer, that is, the ability to add traffic to and remove traffic from a high-speed stream of digital signals. A DCS contains many other functions as well.

The *grooming function* supports efficient use of a given facility. Grooming is the process of organizing and assigning traffic routes to facilities that makes the best use of the facilities available. Both incoming and outgoing signals are groomed. The cross-connection function of a DCS connects tributaries to other tributaries or to higher-speed facilities. Consolidation and segregation are a part of grooming.

The DCS uses a *consolidating function* to better utilize the network. The DCS can improve the fill factor of a given facility by combining low-speed, partially filled incoming facilities into a smaller number of more highly utilized outgoing facilities. Thus, a DCS can improve the cost effectiveness of high-speed facilities.

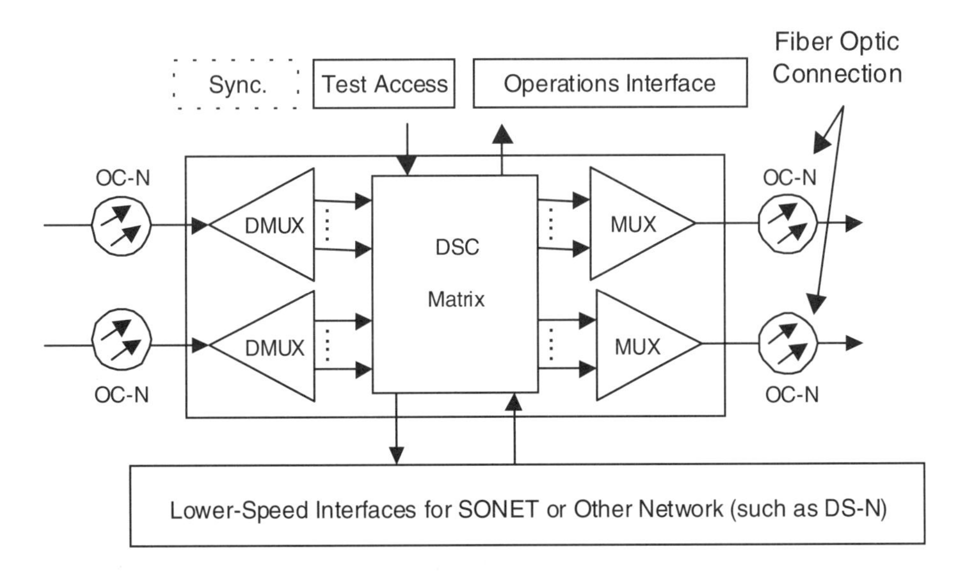

Digital Cross-Connect System Functional Architecture
Fig. 14-11

The *segregation function* of a DCS can support more efficient maintenance when mixed low-speed, incoming facilities are sorted out by service type, destination or protection category. This sorting process is called segregation. Like types are combined, so outgoing facilities can contain a uniform content and are, thus, more efficient to maintain.

Some DCSs can provide a *broadcast function* which supports a one-way cross-connection from an incoming digital signal source to more than one outgoing interface port, or to more than one tributary. This is called a 1-to-N, one-way cross-connect.

A DCS supports a *facility rolling function* which permits transfer of traffic from one facility to another without service interruption. This function reduces the operator's cost and time for connection provisioning. DCSs automate facility rolling to a large extent. Facility rolling can be performed on uni-directional or bi-directional traffic. The facility rolling function of a DCS can support transport facility upgrades of digital technology or digital switch cutovers without service interruption.

DCSs can support a *facility performance monitoring function* to detect performance problems of the transmission paths that pass through the system. Maintenance is more effective when a DCS provides performance parameter counts, as well as threshold and derived parameter values. These counts and values can be analyzed by maintenance systems or personnel and then applied to focus maintenance resources in a more concentrated and rapid manner.

A *monitoring test access function* is supported on a DCS. One mode of test access is non-intrusive testing which allows maintenance personnel to perform tests without interrupting existing traffic or removing facilities from service. Another mode of test access available is the split-test access. This test access permits equipment to be connected to perform more extensive out-of-service tests on a path. Both of these test access arrangements can improve maintenance effectiveness through more efficient dispatch of maintenance personnel and improved prioritization of trouble conditions in the network.

14.3.2 Digital Cross-connect System in Topology Applications

A DCS can apply a variety of its functions depending on where it is placed in the network. For example, DCSs can be used in gateway, hub, mesh and ring topologies. See Figure 14-12 for a basic diagram of DSC deployment examples in these topologies.

A SONET DCS can provide a *transport gateway* (or interconnection) between existing DSn networks and SONET networks. This could assist a carrier in making the transition to a SONET network by preserving its investment in an installed base.

A DCS can also provide an *operations gateway* between a Network Element and an Operations System (OS) that supports those NEs. Figure 14-12 shows a DCS providing an operations gateway for NEs connected on and through a ring topology to an OS. The OS then provides routing and processing functions to support efficient operation and maintenance of the network.

A major function of a DCS is to provide cross-connection of traffic from multiple locations and distribute traffic to other designated locations. A DCS can do this

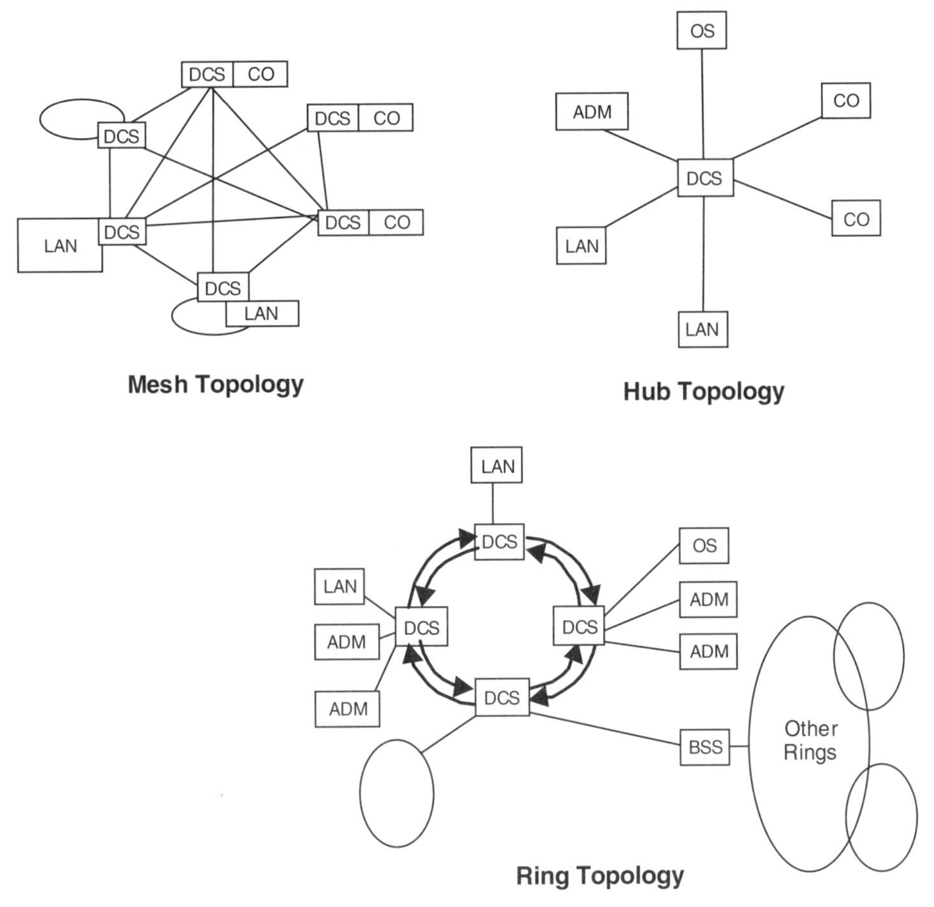

DSC Topology Applications
Fig. 14-12

through a *hub* topology where each traffic generating point (such as a central office or LAN) has a direct link to a DCS at the hub location. The DCS also provides automated network management capabilities for the hub arrangement.

A SONET DCS can perform cross-connection, dynamic traffic routing and network reconfiguration in a *mesh* topology. This function is valuable to support engineered quality of service levels under variable traffic loads.

A SONET DCS can be used between *rings* without being part of the ring. Here, a DCS provides grooming and consolation functions for traffic on a combination of the rings, leading to more efficient facility use.

Ring, mesh and hub topologies are discussed further in Section 15.

In addition, a SONET DCS can perform operations to support applications such as bandwidth management and network restoration.

• *Bandwidth management.* The DCS can automatically process end-to-end high bandwidth connection and disconnection requests. This is a powerful provisioning tool that supports high-speed digital services, which rely on variable bandwidth.

- *Network restoration.* The DCS can reroute circuits in times of network failure. For example, a DCS supports network restoration by moving traffic to another facility port in a hub topology in case of failure. A DCS also supports traffic reconfiguration over multiple alternate routes in a mesh topology in cases of a network failure.

14.3.3 Wideband Digital Cross-connect Systems

One type of DCS is the Wideband DCS (W-DCS). It cross-connects signals at 1.728 Mbps for SONET signals, and/or at DS1 rate of 1.544 Mbps. The W-DCS provides a wide range of interconnection functions for cross-connecting various speeds of SONET circuits. For example, the W-DCS accommodates the DS1 and lower level signals which connect to many more Network Elements and lower-speed applications. The W-DCS is able to look at lower speeds by examining a lower level of the SONET signaling protocol overhead. This lower speed flexibility provides more circumstances for a carrier to apply the DCS network functions (such as grooming or monitoring).

In addition to the general functions listed above, the SONET W-DCS supports several specific functions, including:

- A *gateway* for DS3/DS1/OC-N/STS-N, where OC-N and STS-N are signals "n" times the basic rate of the OC-1 and STS-1 SONET signal (51.84 Mbps).
- *Grooming* for DS3/DS1/OC-N/STS-N
- *Test Access* for DS1
- *Replacement* for a DCS3/1
- *Performance Management* using SONET overhead

14.3.4 Broadband Digital Cross-connect System

The second type of DCS is the Broadband DCS (B-DCS) which typically cross connects SONET signals at the DS3 and STS-N signal rates. This cross-connect system is designed for higher speed interfaces, and it operates in a more efficient mode by looking at a higher level in the SONET signaling overhead protocol.

The SONET B-DCS provides several specific functions. In addition to the general functions listed above, it provides:

- A gateway for DS3/OC-N/STS-Nc, where "c" denotes concatenated signals. Concatenated signals have been added together in a special way to transmit larger payloads to be transported together.
- Grooming for DS3/OC-N/STS-Nc
- Test Access for DS3
- Replacement for a DCS3/3
- Performance management using SONET Overheads.

Digital Cross-connect Systems, whether W-DCS or B-DCS, can support cross-connects with or with out a multiplexing function. For example, one type of B-DCS can terminate a DS3 signal and cross-connect at the DS3 line rate without the multiplexing function. Another type of W-DCS terminates and multiplexes incoming DS3 signals and cross-connects at the DS1 level, performing a multiplexing function.

14.4 NEXT GENERATION DIGITAL LOOP CARRIER

Next Generation Digital Loop Carriers (DLCs) support the integration of DS0 traffic into SONET arrangements. DLC applications include DS1 level transport, which is the current commonly deployed DLC transport level. As the DS0 level of the DLC is included in SONET arrangements, the DLC will become a more powerful and efficient Network Element.

With the DS0 level in SONET arrangements, the DLC can provide a major SONET link directly to the millions of outside plant loops (access lines) in the telephony network. This direct link carries the speed, efficiency and quality attributes of SONET and ATM further into the network and more end users.

Two types of Digital Loop Carrier systems are typical in a SONET system. These are the Universal DLC (UDLC) and the Integrated DLC (IDLC).

The UDLC consists of Remote Digital Terminals (RDTs) and Central Office Terminals (COTs). The COT places traffic from the Central Office onto the digitized local loop for transport to the far end. The RDT receives and terminates the traffic from the network at the far-end of the loop.

In an IDLC system, the RDT communicates with the Integrated Digital Terminal (IDT) of the switch, and the SONET-based RDT-IDT traffic is transported using byte-synchronous DS1 mapping. The IDLC's RDT OC-N interface communicates directly with the IDT's OC-N interface eliminating the need for a COT.

14.5 SONET COMMON NE OPERATIONS FEATURES

All SONET NEs (including the ADM and DCS) include Operations, Administration, Maintenance and Provisioning (OAM&P) capabilities to support the overall SONET system design. This section will highlight many of the functions common to NEs that support SONET OAM&P. Section 17 will discuss how these NE functions are integrated with Operations Systems and other parts of SONET to provide overall efficient high-speed transport at the desired quality level.

14.5.1 Memory Administration

SONET NEs include memory administration capabilities to control and administer their databases. The functions of memory administration include data manipulation, memory backup and restoration, and system administration (including security). SONET NEs have the capability to generate and process certain bits of information to support these functions. These capabilities are important for the overall support of OAM&P of the SONET system.

The memory administration data in a SONET NE includes termination, cross-connection and/or multiplexing configuration information. This also can include information specific to a particular signal, such as the mapping method of a digital signal payload (e.g., asynchronous or byte synchronous).

A SONET NE can be administered by a *data manipulation* function, for example, in order to provision new services. Data manipulation functions include entering, editing, deleting and retrieving data in the database of the NE. Where this function is equipped in an NE and connected to provisioning Operations Systems, data manipulation can be done remotely, saving expense in the service provisioning process.

A *memory backup* function is critical to a SONET NE to support the overall QoS on the SONET system and to maintain the system efficiently. Memory backup in a NE includes local, primary and nonvolatile (or stable) memory backup. This backup scheme is designed in multiple stages to reduce restoration time and the possibility of lost data during a failure. For example, all SONET NE data is backed up automatically in at least one nonvolatile backup memory module after each primary memory update.

SONET NEs are designed to have their configurable memory reestablished by a remote memory *restoration* function identified by the network provider. This function also supports the overall QoS on the SONET system and system maintenance efficiency. SONET NE restoration can be accomplished easily where the NE has the ability to have its nonvolatile memory backup restored by bulk file transfer methods or by a remote memory restoration management application (such as from an Operations System or controller).

System administration and security for SONET NEs are housekeeping functions needed for proper operation. Administration functions include setting the date, time and NE identification. Security functions include establishing routing functions (within the control network), logins, passwords and security levels (including screening options). These security features can prevent unauthorized communications to the NE via any ports and communications channels accepting operations-related command inputs.

14.5.2 Maintenance

SONET NEs support a robust maintenance process. NE maintenance functions include alarm surveillance, Performance Monitoring (PM), testing and control features. Since SONET NEs have many common functions they can be effectively integrated into a service provider's overall effective maintenance program.

This section describes the common maintenance functions contained in SONET. Section 17 of this book discusses how these NE features are integrated with other SONET system features for the overall maintenance process. Common tasks that the NE maintenance capabilities support include:

- *Trouble detection* to identify defects and declare failures. As failures are declared by the NE, that information can be used by the NE itself and/or transmitted to other NEs and Operations Systems within the network to take appropriate action.
- *Trouble or repair verification* to confirm the continued existence or nonexistence of a problem before beginning or closing out work on that problem. This task is important to maintain an efficient maintenance program so that trouble conditions are not ignored or given attention after they have been cleared.
- *Trouble sectionalization* to identify the failure at one of the terminating NEs or the facility that connects them. Sectionalization divides and redivides potential trouble areas into smaller and smaller sections to quickly identify

the exact problem area in the network. Alarms, maintenance signals, Performance Monitoring data, test access and loopbacks within the NE are used to sectionalize trouble in the network. An effective maintenance process by a carrier or operator of a SONET system uses NE sectionalization (as well as other tools and processes) to quickly apply resources to the exact site of the trouble.

- *Trouble isolation* to locate failures down to a replaceable circuit pack, module or fiber. Trouble isolation is a process used after sectionalization to refine the trouble site further. After isolation, the maintenance technician can be more certain of the part which needs to be replaced. NE test access, loopbacks, performance data and diagnostics are used to attain this isolation.
- *Restoration* to resume service even though the failure may not have been repaired. Restoration can be initiated automatically by an NE. For example, protection switching or rerouting traffic can be used to restore service before the failure is repaired.

14.5.2.1 Alarm Surveillance

SONET NEs can detect certain defects on the incoming signals relevant to the layers of functionality they provide. A defect is defined as a limited interruption in the NE to perform a required function. When a defect persists for a prescribed time (or soaking interval), a failure is declared and a failure indication is set. The failure indication may or may not be automatically reported to the Operations System depending on thresholds (or limits) set by the network provider. Failure indications are retrievable by the OS or from other interfaces in the system. Some failure indications are displayed as local or remote alarm indications. Alarms are categorized into critical, major and minor, depending on the severity and service impact of the failure indication. The network provider sets threshold levels and alarm displays.

14.5.2.2 Network Elements: Other Maintenance Support Functions

NEs have other functions and capabilities to support the overall SONET system maintenance. NE support functions are discussed here and in Section 17. Other NE maintenance support functions include the ability to:

- Generate and process certain SONET overhead bits or octets. SONET uses these bits in the calculation and determination of trouble conditions.
- Detect and clear equipment failures. SONET NEs can detect and clear a limited set of localized equipment failures.
- Detect and terminate defects in the NE itself or in certain interconnected portions of the network.
- Insert, pass and remove maintenance signals. SONET passes maintenance signals between NEs and between NEs and OSs.
- Support Performance Monitoring (PM). Performance Monitoring is the process of continually checking the health and effectiveness of the net work. This process has a number of aspects, such as fault detection, alarm thresholds and alarm indications. SONET NEs are designed to support PM and are part of the backbone of maintenance information and function that

make PM possible.

- Perform diagnostics. Diagnostic tests determine what defective elements or functions, if any, exist on equipment or software. SONET NEs can perform diagnostics based on remote or local commands. In addition, an NE can perform a limited set of diagnostics on itself.
- Perform loopbacks. A SONET NE can loop back a circuit, based on a remote or local command, so that a specific function of the NE, facility or circuit can be checked by maintenance personnel. Loopbacks are one of the functions used in trouble sectionalization and isolation.

For SONET to maintain the selected quality of network and service design intended by its engineers, the NEs, topology, synchronization, Operations Systems and maintenance personnel need to be coordinated by design and implementation. The SONET Operations Section 17 discusses how these elements work together to achieve the desired outcome.

15 SONET TOPOLOGY

15.1 SONET TOPOLOGY TYPES

Part II of this book has examined the SONET Network Elements. Connecting these NEs into cooperative interworking arrangements is done through one or more SONET topologies. The first part of this section examines generic topology arrangements and specific SONET topologies. The second part of this section provides an overview of three methods to combine ATM and SONET traffic over SONET Physical layer facilities, using the ring topology as the example facility for the discussion.

SONET topologies can be configured several ways, including those shown in Figure 15-1. SONET Network Elements are designed to be flexible so that they could be used in these topologies. Some of this flexibility existed in NEs that were adapted for SONET, such as ADMs. However, NEs used in SONET also have been arranged to add survivability and management features, which, when coupled with robust topologies and signaling protocols, make SONET effective and efficient. Various parts of the generic topologies existed in telecommunications and other networks before SONET was implemented. SONET has adapted and improved on the strengths of these existing topologies, combining with them other NEs, signal formats, synchronization and operations processes that make up this robust, survivable high-speed transport system.

SONET topologies are discussed below:

Linear topologies can be used, for example, with ADMs to eliminate the need for back-to-back multiplexing. The SONET linear chain is a simple linear arrangement of SONET NEs, (in this example, ADMs) that use SONET framing and timing synchronization. The ends of the linear chain do not connect with one another. The linear chain can be used in serial point-to-point applications, such as connecting several central offices and/or customer locations together with intermediate ADMs in a series to transport moderate traffic needs.

Hub topologies are used in light point-to-point traffic situations. In a hub topology, each traffic-generating point (e.g., central office or LAN) has a direct link to the hub location. As discussed earlier in the DCS Network Element Section, the hub node can be a DCS. As the traffic reaches

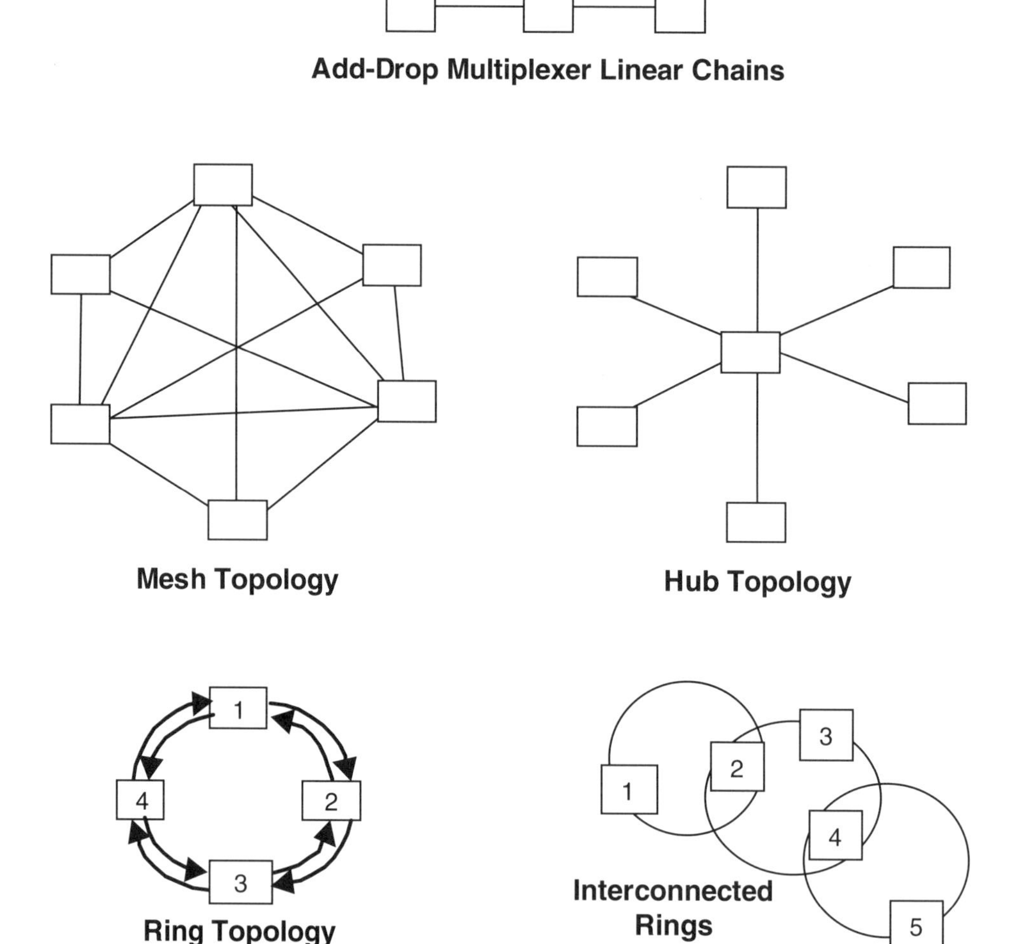

SONET Topology Examples
Fig. 15-1

the hub, the DCS can efficiently connect the traffic to the appropriate node. However, if the link between the node and hub is severed, the transport is lost for that node. If alternate routing (such as a different physical path) is designed between the peripheral node and hub, traffic is split between the routes and the impact of a path cut can be minimized.

Mesh topologies are more effective for heavy point-to-point demands. In a mesh topology, each traffic-generating point (e.g., a central office or LAN) has a direct (or point-to-point) link to many other points. Protection switching in a point-to-point topology can be achieved when protection fibers between adjacent nodes are established. When a failure occurs, the node can then use a protection path through an adjacent node to the final destination.

In the mesh structure, point-to-point connections are sometimes supplemented by "overflow" or alternate routes for the peak periods so that transport delays are avoided. In this case, a mesh and hub topology could be combined providing the strength of the two structures together. The case of alternate routes (or combined topologies) assumes that nodes on the combined topologies have a traffic switching or intelligent routing capability (such as DCS or central office switch). The intelligence in these switches or routers determines when traffic loads dictate the use of alternate routes and where to direct the alternate routed traffic.

Ring topologies provide very quick recovery from network faults or failures. Ring topologies are common and growing applications in SONET and will be discussed in more detail in Section 15-2.

Interconnected rings can be established by interconnecting DCSs or high-speed ports of the ADM. These rings connect networks to provide a seamless, high-speed end-to-end path that can support highly efficient and reliable high-speed digital services. ADMs provide basic add-and-drop functions between the rings, and the DCS can provide grooming and consolidation functions. Typically, interconnected rings are established in a hierarchical manner (such as Customer, Local and Express) to segregate the types of interfaces and speeds in an ascending order. An example of this is shown in the ATM switching interface discussion in the previous sections, as illustrated in Figure 14-6 and Figure 14-7.

As a practical matter, a carrier's entire network typically comprises a combination of topology types, depending on the traffic demands and forecasts among various points in the network.

15.2 RING TOPOLOGIES

A *SONET Ring* is a physical (fiber optic) transmission arrangement connecting SONET nodes. This is the most popular and widely implemented SONET topology.

The SONET Self-Healing Ring (SHR) is a set of nodes and optical fibers interconnected to form a closed loop that includes other SONET NEs. The SHR rapid recovery function is performed by the NEs at the nodes on the ring. This arrangement provides one of the major features of many SONET implementations: the quick recovery from a network failure. The SHR also supports shared facilities to provide this recovery protection from both nodal and facility failures.

Nodes on the SONET rings are typically ADMs, but they also can be DCS or next generation DLCs. The generic topology of fiber optic rings precedes the SONET standard system. It is still used to connect corporate centers on an intra-building or inter-building basis as well as to connect corporate centers to central offices for telephony and data communications. However, this section will focus on SONET rings.

SONET rings exist in these types:

- Uni-directional Path Switched Ring (UPSR)
- Two-fiber Bi-directional Line Switched Ring (BLSR)
- Four-fiber Bi-directional Line Switched Ring (BLSR)

Transmission on the SONET ring topology can be *uni-directional* or *bi-directional*. The SONET ring is multi-fiber (two or four-fiber). Multiple rings can be combined on the same physical path. However, carriers choose the physical path and ring combinations with care to assure as much survivability on the rings as can be achieved within the design and budget constraints of the overall network architecture.

SONET SHRs can be classified as two main types: path switched rings and line switched rings.

In a *path switched ring*, incoming or added traffic is routed in both directions around the ring. Protection switching occurs on a path-by-path basis in this ring when a ring or node failure is detected by a Network Element. Network Elements in this ring use defect indications and maintenance signals in the highest SONET signaling layer (the Path layer) to trigger protection action. Traffic is placed on another path, so different ring nodes are traversed. This supports survivability in the network until the repair is completed.

In a *line switched ring*, incoming traffic is only assigned in one direction unless protection switching has been applied, such as in the case of a network failure. Protection switching occurs in this ring when a ring or node failure occurs. Network Elements in this ring on either side of the failure coordinate to use defect indications and maintenance signals in the middle SONET signaling layer (the Line layer) to trigger protection action, resulting in bi-directional traffic flow, until service is restored.

15.2.1 Uni-directional Path Switched Rings

Uni-directional Path Switched Rings (UPSR) are two-fiber rings. UPSR rings transport the same information (customer traffic) in opposite directions around a ring. Figure 15-2, for example, shows traffic (live path) entering Node 1 and destined for Node 2. In normal operation the traffic would travel directly to Node 2, clockwise around the ring. Live traffic destined for Node 4 is transported clockwise from Node 1 to Node 2 and 3 before reaching Node 4. At the same time, a duplicate copy of the live traffic is being transported in the opposite direction around the ring for protection purposes [34].

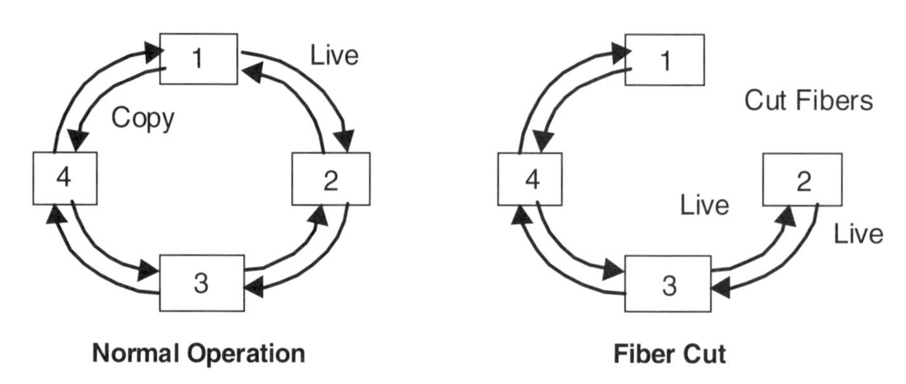

Uni-directional Path Switched Ring
Fig. 15-2

A UPSR ring carries many SONET paths (STS signaling paths) on the ring. A path switched ring topology requires two-fibers to send the traffic in both directions. See Figure 15-2 for a representation of a UPSR.

The UPSR provides protection against single line failures or a single node failure. For example, the diagram in Figure 15-2 shows one working path on the ring. Traffic from Node 1 to 2 is dual fed at the source, Node 1. Live and protection traffic travels around the ring in opposite directions. All traffic on the live fiber travels in one direction and a copy of that traffic traverses the protection fiber in the opposite direction, in normal operation. A cable cut between Nodes 1 and 2, for example, would cause the live path between Nodes 1 to 2 to switch to the incoming copy path (at Node 2) for protection. This means that the live traffic from Node 1 would pass through Nodes 4 and 3 before arriving at Node 2. The returning traffic to Node 1 from Node 2 was not affected by the fiber outage. During the fiber cut (before full service is restored) there is no protection on the ring.

15.2.2 Bi-directional Line Switched Rings

On *Bi-directional Line Switched Rings* (BLSR), the two (transmit and receive) directions of the working circuit travel the same route in opposite directions. Protection switching is performed based on fault and alarm indicators at the Line level (the middle layer of the SONET signaling format), whereas UPSR protection switching is performed at the Path level. A BLSR can be a two- or four-fiber ring.

In a two-fiber BLSR, half the bandwidth is reserved (half the time slots) on each fiber. This is done for protection. For example: If the ring is an OC-48, 24 working circuits could be placed between Nodes 1 and 2. With the 48 time slots available, the 24 working circuits would be on time slots 1-24 and the protection time slots would be placed in time slots 25-48 in the opposite direction around the ring. [35]

This arrangement limits the span capacity of the fiber to OC-N/2, or half the bandwidth of OC-N. However, bandwidth in non-overlapping sections of the ring can be reused.

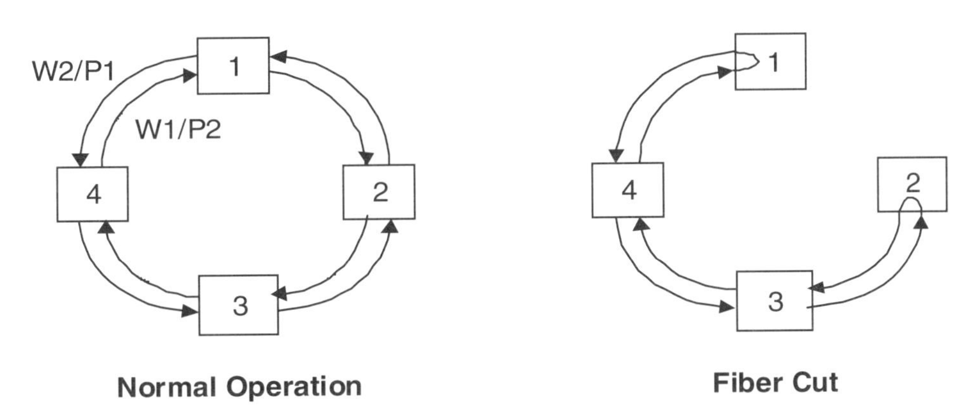

Two-Fiber Bi-Directional Line Switched Ring
Fig. 15-3

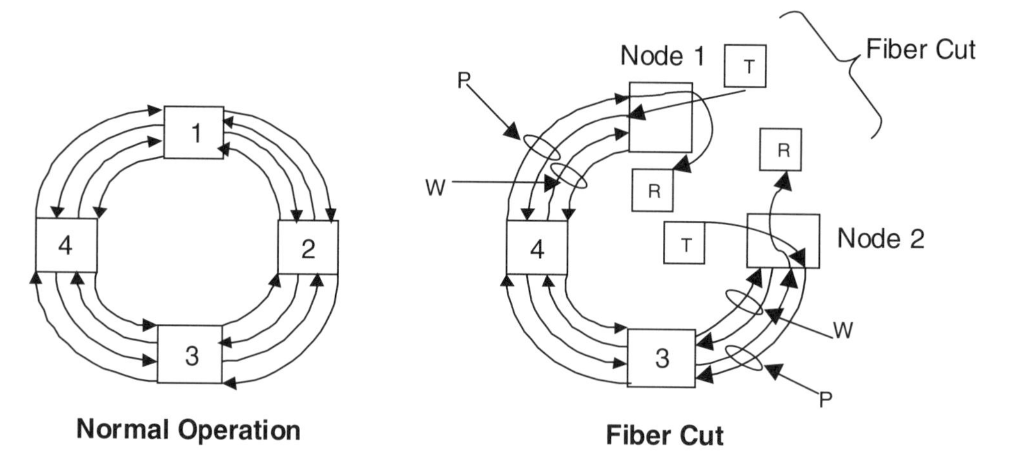

Four-Fiber Bi-Directional Line Switched Ring Using Ring Switch Arrangement
Fig. 15-4

An example of a two-fiber BLSR is shown in Figure 15-3. If the cable were cut between Nodes 1 and 2, the two nodes would perform a ring switch (or loopback) at the Line level. This loopback would connect all working channels (in the node facing the cable cut) to the protection channels in the opposite direction. Thus, a single failure or single node failure will not disrupt communication.

A four-fiber BLSR structure is shown in Figure 15-4. Protection is provided by two separate protection fibers that are in addition to the two working fibers. This arrangement provides a span capacity of a full OC-N. Note that this is twice the capacity of the two-fiber BLSR. In addition, bandwidth in the non-overlapping sections of the ring can be reused.

In this case, a cable cut between Nodes 1 and 2 causes those nodes to perform a ring switch at the Line level. This is shown in Figure 15-4. This is a loopback similar to the one explained in the two-fiber BLSR discussion. The loop back connects the working fibers facing the cable cut to the protection fibers (P) in the opposite direction. In this example, the transmitters (T) for the working pairs (W) in Nodes 1 and 2 transmit on the protection fiber. The receivers (R) in Nodes 1 and 2 receive from the protection fiber. This feature of the BLSR protects against single line failures or single node failures.

An additional protection mechanism is available in the four-fiber BLSR. This arrangement is called a four-fiber *Bi-directional Line Switched Ring* (BLSR)-Span Switch. This can protect against a failure, which affects only the working fibers between adjacent nodes, by placing a span switch between these nodes. In the case of a failure, the working traffic would be switched to the protection fibers between adjacent nodes, as an alternate to the protection fibers in the reverse direction. This is similar to protection switching in a point-to-point or hub topology.

15.3 HYBRID TRAFFIC ON RING TOPOLOGY

Previous sections discussed the transport of hybrid traffic (ATM, SONET and possibly other traffic) over several of the SONET Network Elements. The introduction and growing presence of ATM technology in a carrier's network will generate the need to mix traffic types in the transition to higher levels of ATM traffic across the total network. All the Network Elements are involved in this transition by handling mixed traffic as the network evolves [40].

This section outlines three methods for combining ATM and traditional Synchronous Transfer Mode (STM) traffic (such as DSn and SONET frames). The three methods are examined using SONET rings as the basic facility. Examples of how the other NEs fit in with the ring topology scheme with these traffic mixes also are discussed. Figure 15-5 shows a simple diagram of how the ATM and STM traffic is mixed over a single network.

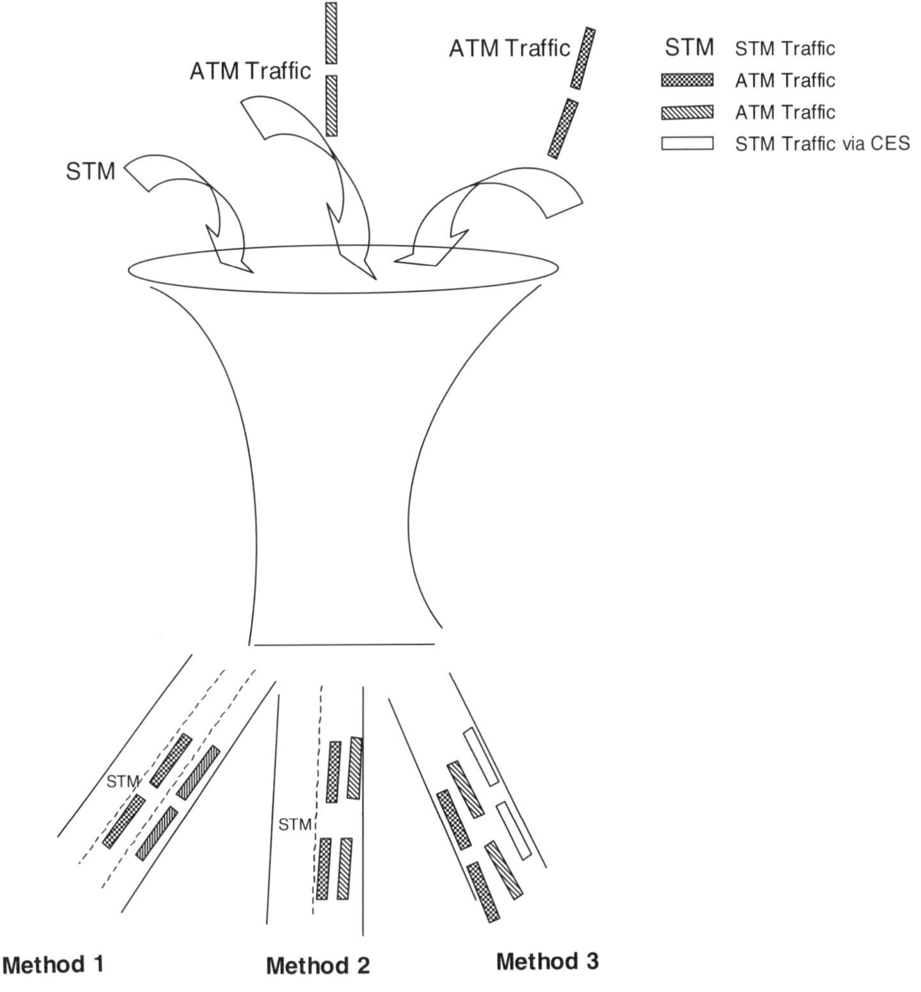

Three Methods of Combining ATM and STM Traffic
Fig. 15-5

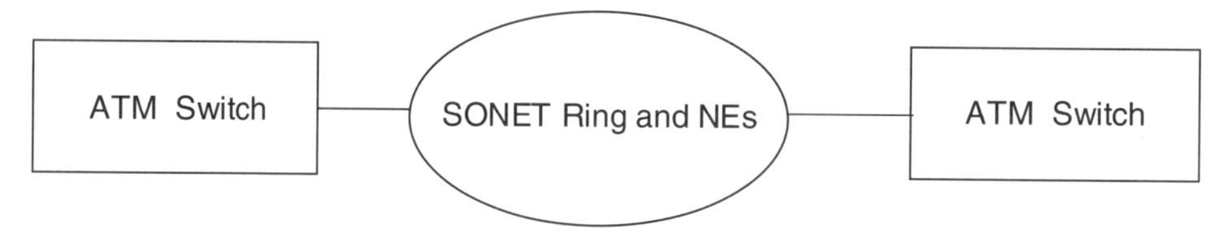

Conventional ATM Switching and SONET
Fig. 15-6

Method 1 for combining ATM and STM traffic uses conventional SONET transport NEs (such as ADMs and rings). This method combines ATM and STM traffic onto the same SONET facility. However, each type and each source of traffic is placed on different SONET transport frames. ATM cells are mapped (or placed on) SONET transport frames, but SONET NEs only process at the SONET layer and do not process at the ATM layer. Each ATM cell (or cell stream source) uses its own SONET Frame, even if only a small portion of the SONET Frame is used. The number of SONET Frames used depends on the overall amount of ATM or STM traffic.

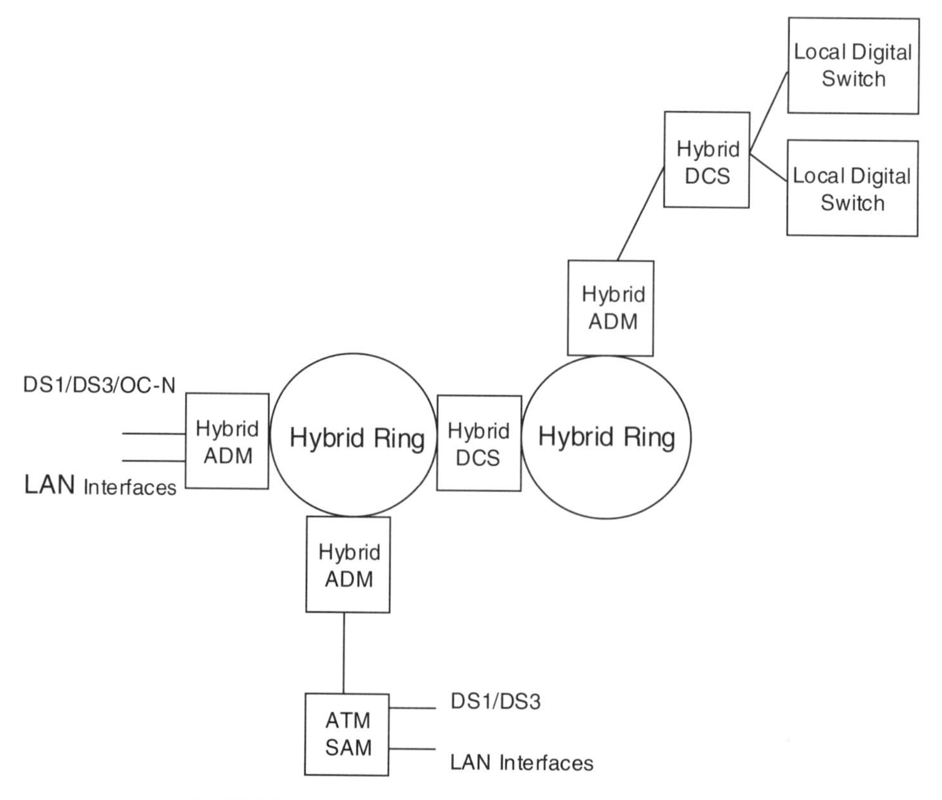

Hybrid SONET ATM NEs Network Including Hybrid Rings
Fig. 15-7

Figure 15-6 shows an example of how ATM switches are used with SONET ring and ADM NEs. ATM edge and hub switches can be used for ATM cell switching and SONET NEs for transport.

Method 2, for combining ATM and STM traffic over a signal transport network, uses hybrid ATM and SONET NEs. This method combines ATM and STM traffic onto the same SONET facility over different SONET transport frames. Method 2 is different from Method 1 in that ATM traffic from different sources can be aggregated and carried over a single SONET transport frame.

Hybrid SONET/ATM means that the network, traffic and NEs can handle (and access) both the SONET and ATM layers of signaling. Figure 15-7 shows an example of a hybrid SONET/ATM network using hybrid NEs. An ATM Service Access Multiplexer is used at the edge of the ATM network to provide ATM interfaces to customer services. The hybrid ring (containing SONET and ATM hybrid NEs) is a SONET based ring. This ring can be either UPSR or BLSR. On this hybrid ring, the ADMs have ATM cell processing capabilities. This allows several ring nodes to share a SONET payload for ATM traffic. The hybrid DCS provides ATM Virtual Path level cross-connect, grooming and management capabilities.

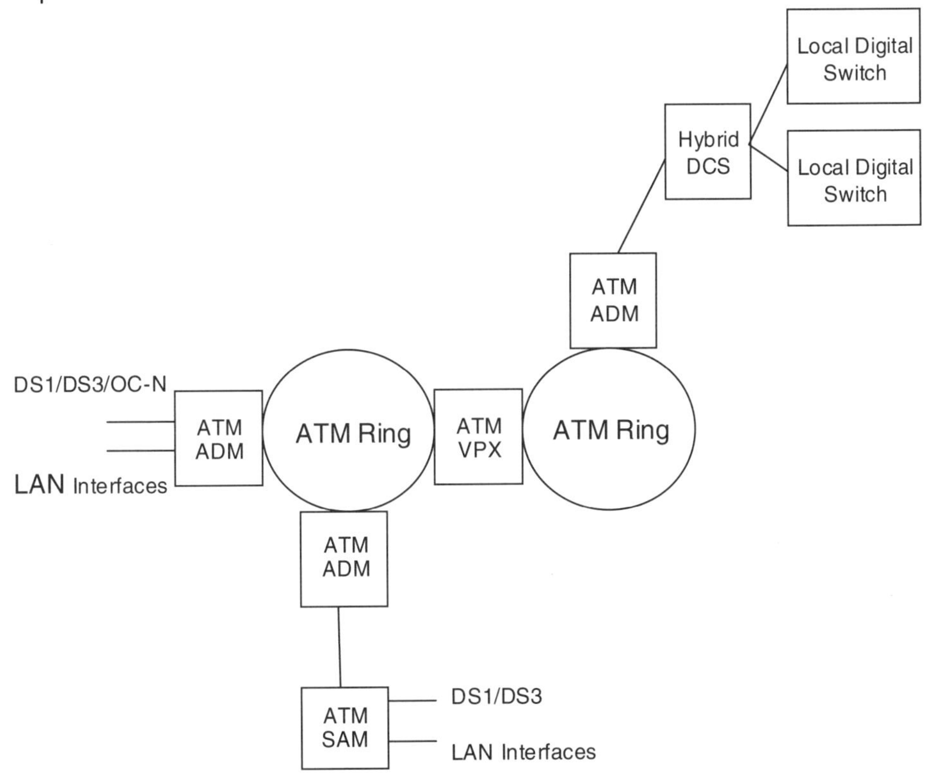

Pure ATM NE Network Including ATM Rings
Fig. 15-8

Method 3 for transporting ATM and STM traffic over a single transport network uses pure ATM NEs that convert the STM traffic into ATM cells (via ATM Circuit Emulation Service). In this method, all STM and ATM traffic is carried as ATM cells in a single concatenated SONET formatted frame. Figure 15-8 shows a pure ATM network using ATM NEs. The NE designated VPX is a cross-connect system operating at the ATM Virtual Path level, not at the SONET signal level where a SONET DCS would operate. This network is similar to the hybrid network discussed in Method 2; however, STM signals are carried throughout the two ATM rings via circuit emulation and restored to STM before they continue their path through the hybrid DCS to the local digital switch.

15.3.1 SONET and ATM Network Elements Classification Comparison

Network Elements are classified depending on their ability to handle different levels of signals and layers of protocol. Section 15 discussed SONET Network Elements. However, as ATM becomes a larger portion of the overall network, NEs that can handle the transition to an ATM environment are needed. As portions of the network become completely ATM, NEs that operate purely at the ATM level protocol will be most efficient. The following classifications have been developed to clarify the engineering and implementation of ATM and SONET systems and transition from existing networks.

SONET NEs can process all traffic at the STS layer (such as time slot processing/routing and time division multiplexing). The SONET NE can also process or transparently pass STS signals (such as DSn and SONET frames) that contain ATM cells as payloads. This is referred to as conventional ATM on SONET. For example, SONET ADMs add and drop a variety of digital signals from a high-speed SONET OC-N signal. The added and dropped signals from the low-speed ports may be SONET framed signals or DSn signals.

ATM NEs can process all traffic at the ATM layer (ATM cell processing/routing and ATM cell multiplexing). These NEs have SONET signal level or other Physical layer interfaces for the transport of ATM cells. At this type of interface, the SONET or other Physical layer is always terminated and the ATM cells are separated from the payload. The ATM NE also may be able to process or transparently pass ATM cells that contain STS signals (such as DSn and SONET frames) in the form of ATM circuit emulation. For example, ATM ADMs are similar to the SONET ADMs. However ATM ADMs add and drop ATM Virtual Path signals at their low-speed ports. ATM VPs are part of the transport capacity that passes between the high-speed ports of the ATM ADM. The high-speed ports of this NE may use SONET frame signals or other Physical layer interfaces for the transport of the ATM cells. At the high-speed interfaces, the SONET and Physical layer signals are always terminated, even if the ATM cells pass through the ADM without being added or dropped.

Hybrid NEs can process the traffic at the STS layer (time slot processing/routing and time division multiplexing) as well as at the ATM layer (ATM cell processing/routing and ATM cell multiplexing). The hybrid NE can perform the functions of a SONET NE as well as those of an ATM NE. Thus, the hybrid ADM is a SONET ADM with ATM cell processing capabilities.

16 SONET TRANSPORT PROTOCOL

As discussed previously, designers have built a robust foundation on which to send efficient high-speed signaling by combining the SONET NEs and topologies. A strong and flexible format and structure is needed for this foundation to achieve the next portion of the SONET system, the transport mechanism.

SONET transport is a progressive hierarchy of optical signal and line rates. It has a unique and standardized frame format and structure, which balance the payload carrying capacity necessary for efficient transport and administrative overhead. The SONET signal administrative overhead supports efficient routing, error correction, monitoring and quality control [36].

16.1 SONET HIERARCHY OVERVIEW

16.1.1 SONET Hierarchy Overview-Frame

SONET transport formats are built by sending, receiving, processing and managing a series of frames (or blocks) that have a defined character and limited number of configurations. This is a major strength of the SONET protocol. SONET has less administrative overhead that other transport methods because there is a finite and predictable set of block configurations. Even with reduced administrative overhead, SONET provides a high level of throughput while establishing a robust framework for management of the system.

The basic building block of SONET transport is a frame called *Synchronous Transport Signal at Level 1* (STS-1) for the electrical format. Higher rates of SONET signals are multiples of STS-1 and are designated as STS-N (where *N* is a whole number) signals. In the optical realm, these signals are OC-1 and OC-N, respectively. These higher-rate signals are developed by byte-interleaving (interweaving signal bytes together in a prescribed manner) "n" multiples of the basic STS-1 signal to achieve higher rates of transmission. The SONET frame and its organization is examined in more detail in the following sections.

16.1.2 SONET Hierarchy Overview-Layers

SONET transport uses a layered approach. This provides an organized and predictable framework for transport as well as system management.

The lowest SONET layer is called the Physical Media Dependent (PMD) layer. The PMD layer provides for the optical or electrical transmission of bits. SONET Network Elements (NEs) (for example the Terminal configuration ADMs) provide termination at least at the PMD layer.

In the SONET hierarchy, subsequent layers are logical layers (Section, Line and Path layers, in that order). These layers are based on the seven-layer Open System Interconnection (OSI) Reference Model. All SONET layers are sublayers of the lowest OSI layer, called the Physical layer. Every SONET logical layer has associated overhead that is generated and terminated by NEs operating at that level. The NEs that generate and terminate SONET layers are called Section Terminating Equipment (STE), Line Terminating Equipment (LTE) and Path Terminating Equipment (PTE), in ascending order. The SONET NE may terminate at one or more layer. For example, the STE, LTE and PTE functions may be combined in one piece of equipment, such as an ADM.

The SONET frame format and layers of its organization are discussed in more depth below.

16.2 SONET FRAME

16.2.1 STS Frame-Introduction

As with other high-speed transport mechanisms, SONET Network Elements transmit information serially, that is, one bit at a time, one after another, using binary coding techniques. However, in SONET, the bit streams are segmented into an 8-bit octet (or byte). A group of 810 octets form a *Synchronous Transport Signal at Level 1*, or STS-1 frame. The STS-1 frame is the basic building block or fundamental signal in SONET. The structure of STS-1, which includes a scalable transmission rate and administrative overhead, is designed to fit together with the NEs, topologies and Operations Systems of SONET to form a seamless and efficient high-speed transport system.

The SONET transport system is designed to interface with existing and lower-speed elements and networks. One mechanism that facilitates that interface is the SONET frame. The SONET frame converts the STS-N electrical signals to the SONET OC-N optical signals so that network elements based on electrical signals (such as traditional wireline or wireless telephony switches) can be interconnected in a high-speed and highly efficient manner over a fiber optic facility. The STS-1 is the electrical equivalent of an optical OC-1. Thus, the basic rate of SONET can be expressed as STS-1 for signals in the electrical format or OC-1 for signals in the optical format.

16.2.2 STS Frame-Structure

The STS-1 frame consists of the overhead and the payload. The overhead portion of the SONET frame ensures the signal integrity and support organization of the signal payload. The overhead portion of the STS-1 frame is called the *Transport Overhead* (TOH). The payload portion of the STS-1 frame is called the *Synchronous Payload Envelope* (SPE). The STS-1 SPE contains the content data to be transported. For flexibility and efficiency, the space in the STS-1 SPE is organized into *payload, Path Overhea*d (POH), *Virtual Tributaries*

(VTs) and *Virtual Tributary Overhead* (VTOH, if VTs are used). The functions of the various portions of the STS-1 frame are:

- STS-1 SPE Data or payload storage location within the STS-1.
- TOH STS-1 overhead responsible for ensuring the integrity of the overall transport of the frame.
- POH STS-1 overhead responsible for ensuring the integrity of SPE payload.
- VT SONET structure for organizing smaller payloads within the STS-1.
- VTOH Overhead responsible for ensuring the integrity of individual payloads where a VT is used within the STS-1.

16.2.3 SONET Frame-Frame Rate

STS-1 is the basic building block of SONET and is the SONET standard for transmission over optical fiber in the optical format, OC-1, at the rate of 51.84 Mbps. The STS Frame is sent through in a row-by-row transmission, starting with the most significant bit. Frames are transmitted from left to right. Each frame in the STS-1 is transmitted in 125 microseconds (μs).

The basic transmission rate of STS (51.84 Mbps) is derived from the total of the bits in the STS-1 frame format which is delivered in 125 μs. For example, the OC-1 optical frame consists of a format of 9 rows by 90 octets, which is transmitted at the rate of 8000 frames per second [36].

$$(9 \times 90)\frac{\text{bytes}}{\text{frame}} \times 8\ \frac{\text{bits}}{\text{byte}} \times 8000\ \frac{\text{frames}}{\text{second}} = 51{,}840{,}000\ \frac{\text{bits}}{\text{second}}$$

$$= \mathbf{51.840\ Mbps}$$

16.2.4 SONET Frame-Higher Rates

STS-N is the Synchronous Transmission Signal rate of the electrical signals. The "N" is the index of *n* times the basic rate 51.84 Mbps (i.e., *n* times STS-1). Currently, recognized values by standards bodies of "n" multiples of the STS-1 signal transmission rate include 1, 3, 12, 24, 48 or 192. OC-N[13] is the SONET transmission rate of the optical signals. "N" rate multiples are the same for STS-N electrical and OC-N optical formats. For example, OC-3 is 155.52 Mbps. This is the same rate as STS-3. However, it must be understood that optical and electrical signals are of a different type and the "n" refers to a neutral index.

Common SONET rates used to transport ATM, expressed here in the optical format, include OC-3, OC-12 and OC-48. OC-24 is not used. Higher rate transmission at OC-192 (10 Gigabit) and Wave Division Multiplexing (WDM) interfaces are in the early phases of development and standardization in the industry.

When SONET system designers (through the industry standards process) considered how to establish SONET frame formats and rates, they had in mind

[13] The designation of lower case "c" after OC-N, for example, OC-3c, indicates that the channels or signals are *concatenated.* That means they operate as a single signal channel rather than as *n* multiplexed independent channels.

the goal of developing one global fiber standard. To do this, designers had to develop a frame structure and rate that would allow all the different existing transmission hierarchies around the world to be placed into one common standard format. The SONET/Synchronous Digital Hierarchy (such as SDH)[14] standards emerged. For example, the basic OC-3 rate allows the multiple DS1/DS3 and multiple E1/E3 transmission systems to map into the SONET/SDH standards.

16.2.5 SONET Frame-Frame Format

The SONET, frame format structure was developed to efficiently place existing transmission schemes (such as electrical signal formats) into the SONET system. The SONET frame format contains overhead and flexible payload arrangements to accommodate many of these existing transmission schemes.

SONET designers added a small amount of overhead to more efficiently carry electrical signal formats over fiber optic technology. This overhead consists of transport and path overhead for synchronization, error management and transmission management functions.

In addition, SONET accommodates small and large payloads. This payload flexibility can support services as diverse as a short, bursty messaging, full text messaging, voice and video transmission. Figure 16-1 provides a pictorial overview of the STS-1 frame format.

In the STS-1 frame structure, each row of the SONET frame format consists of a three-byte (or column) Transport Overhead (TOH) and an 87-byte Synchronous Payload Envelope (SPE). The TOH is divided into Section Overhead (SOH) and Line Overhead (LOH) portions. The SPE divided into a one-byte path overhead and the payload capacity.

16.2.5.1 STS Transport Overhead

As shown in Figure 16-1, the first three bytes of the STS-1 frame are the *Transport Overhead* (TOH). The TOH is used, as the name implies, to transport the STS-1 across the medium. The Section and Line layer equipment insert the TOH as a part of the overall transport and administration of the SONET frame.

The TOH structure consists of 27 bytes. Nine bytes (three rows) are overhead for the Section layer (SOH), and 18 bytes (six rows) are for the Line layer (LOH). The Section and Line layers are two of the SONET Management Layers (described in Section 17). The TOH contains information relating to the transport and administration of the SONET system, such as STS-1 framing, error monitoring, error detection and support of Automatic Protection Switching. The TOH is modified or created by the Network Element that processes (or terminates frames from an incoming interface and inserts them into the outgoing signal) at the Section or Line layers of SONET.

16.2.5.2 STS Payload Envelope

The STS Payload Envelope consists of the Path Overhead and the payload itself.

---

[14] Section 19 discusses the differences between the SONET and SDH systems. As industry committees continue to work on these systems, the differences become fewer over time.

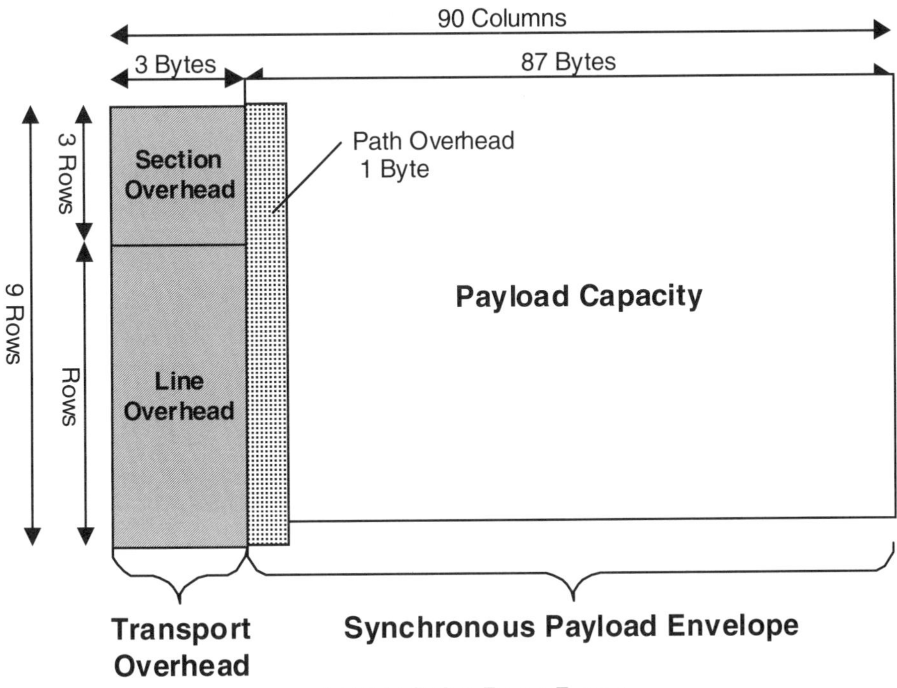

SONET STS-1 Frame Format
Fig. 16-1

STS Path Overhead Structure

The STS *Path Overhead* (POH) is associated with each payload. It communicates information regarding the point at which the payload is mapped into the STS-1 SPE and where it is to be delivered. Also, the POH is the overhead which is responsible for ensuring the integrity of the SPE payload and includes transmission error checks. The POH occupies the first column (byte) of the STS-1 SPE.

16.2.5.3 STS Payload

The SONET payload structure can accommodate asynchronous and synchronous signals. In accommodating asynchronous signals, SONET takes a major step in placing existing transmission schemes into its structure. The process of placing either of these type signals into a SONET payload is called *"mapping"*. Once these signals are mapped into SONET, however, they become synchronous signals.

Mapping refers to the type of packaging structure a payload will use for transport within a SONET signal. These packaging structures are different for different payload signals. For example, a DS3 signal will use a different mapping structure than a DS1 signal.

The asynchronous DS3 mapping structure example, shown in Figure 16-2, uses two columns of fixed stuff bytes (columns 30 and 59) to bring the DS3 signal to the *approximate* bit rate of the STS-1 SPE. To reach the *exact* bit rate

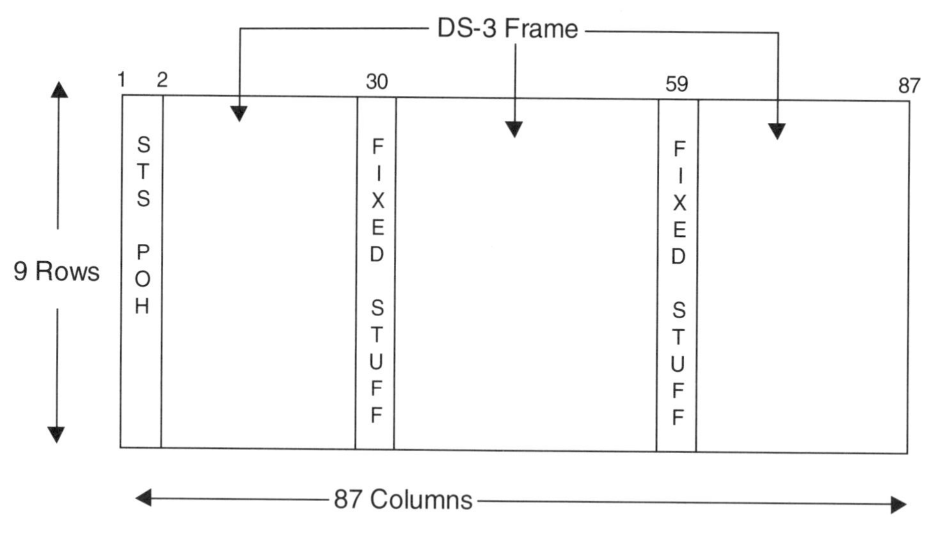

SONET - DS3 Asynchronous Payload Mapping Structure
Fig. 16-2

of the STS-1 SPE, a DS3 single bit stuffing mechanism is used. Stuff bytes and bits are extracted at the far end (the end of the network where the signal exits the SONET system) before the intact DS3 is passed on to the other asynchronous DS3 equipment, such as a M13 type multiplexer.

The most common payload sizes carried over SONET networks are the DS1 and DS3. For example, an OC-1 optical format signal could carry 28 DS1s or 1 DS3. SONET payload size levels increase at whole number multiples and, thus, so do their capacities.

In another payload mapping example, placing ATM signals into an STS payload requires that the ATM cell crosses the Synchronous Payload Envelope boundary and is converted to STS-3c. This is necessary because the ATM cell rate (149.76 Mbps at STS-3c rate) is not an integer multiple of the 53-byte ATM cell. In this example, SONET demonstrates the flexibility to carry payloads that are split across payload envelopes. This flexible capability is important to support high-speed services, which are less (or not at all) tolerant of delays or significant errors (such as video).

One of the ways for SONET to achieve its high efficiency at a high speed is a flexible frame format. The payload capacity, or SPE, can start anywhere in the STS-N Payload Envelope. For example, the first column of the SPE may not coincide with the first column of the Payload Envelope, as shown in Figure 16-1. In order to maintain this flexibility, a system of pointers is needed to indicate where the SPE starts [37]. The next section will discuss these pointers in STS.

16.2.5.4 STS Pointers

Conventional Multiplexer Mapping

In conventional multiplexing, fixed location mapping can accommodate large, asynchronous frequency variations, but access to payloads requires destuffing and identifying of the frame pattern of the payload. (Destuffing is the process of separating useful bits from "dummy" bits.) Fixed location mapping requires 125 μs buffers to phase-align, delay and slip the tributary signal. For this reason, an improvement in conventional multiplexing is desirable. SONET was developed, in part, to improve the conventional mapping scheme.

STS Payload-Pointer Structure

In SONET, every STS-1 frame carries a payload pointer in the overhead portion of the signal. This payload pointer is an important element for achieving improvements in multiplexing. SONET pointers improve multiplexing over conventional methods by assuring easy access to synchronous payloads and, at the same time, avoiding the need for 125 μs buffers and their associated problems. Through extensive use of pointers, both asynchronous and synchronous payloads can be carried in a synchronous payload.

STS Synchronous Payload Envelopes (SPEs) are not required to be aligned, for example when a payload overlaps into other STS frames. This is because every STS-1 has a Payload Pointer to indicate the location of the SPE or to indicate concatenation.

A *pointer* is a number (or address) carried in the STS-1 line overhead that indicates the starting location of the STS-1 SPE payloads within the STS-1 frame. Thus, the payload is not locked into the STS-1 frame but can float with respect to the STS-1 frame. The pointer also provides a synchronization function in SONET.

In STS, a pointer indicates the start of the SPE. In the STS Transport Overhead, pointers are labeled as bytes H1 and H2. The H3 pointer adjustment byte and the 0 byte of the SPE overhead manages the pointer adjustment, which helps keep SONET synchronized. Typically, the Payload Envelope begins in one STS-1 frame and ends in the next.

Because the SPE is not locked to the STS-1 frame structure, the floating SPE can be shifted to account for frequency deviation in the transport of the payload envelope. SPE pointers track the frequency deviations, or shifts. The differences can be accumulated in a buffer until a threshold is reached. Once the threshold level is reached, an adjustment is triggered. These pointer adjustments occur when the incoming signals are faster or slower than the output clocking circuitry. Generally, Network Elements are designed to continue processing functions when pointer adjustments occur less often than every 500μs (or once in every four STS-1 frames). These timing deviations, which could require adjustment, can be the result of:

- Noise
- Clock Transients
- Interruptions in Facilities
- Network Element Deviations
- Other Environmental Factors

16.2.5.5 STS-N Frame Structure - Larger Payloads

When SONET needs more than one frame to transport payloads, an organized and consistent structure is needed to multiply the carrying capacity of the

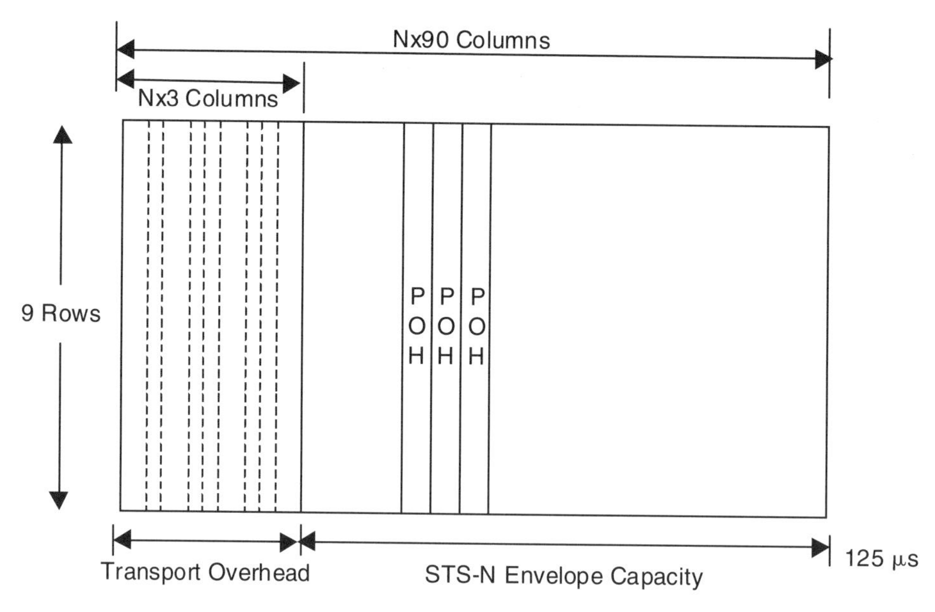

SONET Frame Structure STS-N Frame (Example N = 3)
Fig. 16-3

single STS-1 frame. The SONET frame structure for *n* frames is depicted as the structure of an STS-N sequence of *n* x 810 octets. This is shown in Figure 16-3, wherein "n" is 3.

For example, in order to transport three STS frames, each field of the STS-N frame format is *n* times the octets in each row. In this example, the payloads are not necessarily aligned. For this multiple of *n*, STS-3 has 9 octets of Section overhead, 9 octets of Line overhead and a 261-octet payload envelope in each row with each column being 9 octets in length. Thus, by interleaving three STS-1 modules, the STS-3 structure is formed and provides a specific sequence of *n*x810 (9x90) octets. This structure provides 2430 (3x810) octets for STS-3 transport.

As the STS frame structure is expanded beyond *n*=1, and before interleaving, the Transport Overhead of each of the STS-1 modules is frame and phase-aligned. STS pointers support the frame and phase alignment of the STS-N signals. Pointer offsets are calculated for any needed adjustments. The pointer-offset adjustments support minimum buffering and signal delay in the STS-1 transport.

Payload pointers support alignment functions by decoupling the SPE frame alignment from the STS-1 frame alignment, which is why the SPE floats with respect to the STS-1 frame. This enables multiplexing synchronization in an asynchronous environment and avoids the need for buffering, the resulting delays and the loss or duplication of bytes.

Large, or Super Rate, STS payloads can be transported where alignment is required (for example, B-ISDN ATM applications mapping ATM cells at 149.76 Mbps). The process of linking this type of payload for transport is called *concatenation.* With concatenation, the first column of the SPE contains only one set of STS Path Overhead. For example, in a STS-3c SPE, the SONET frame contains 261 byte columns (3x87) with 9 rows and one set of STS Path Overhead. The other columns, which would have otherwise been used for path overhead, can be used as payload in the case of a concatenated signal.

Other examples of concatenation are STS-12c and STS-48c. Broadband signals that require concatenation include IP, ATM, FDDI and SMDS.

These steps establish the basic elements of concatenation:

- Form an STS-Nc (where "c" denotes concatenation) module. This formation occurs when N constituent STS-1s are linked together in fixed phase alignment.
- Map a Super Rate payload into the resultant STS-Nc SPE for transport. Then, OC-N, STS-N electrical, or higher rate signals can carry the STS-Nc SPE.

Note: The NE functionality will affect the ability of the network to transport Super Rate payloads. Further information on capabilities of various NEs to handle Super Rate payloads can be found in Reference [33]. In one example, a NE supports the multiplexing, switching or transport of STS-Nc SPEs, treating each STS-Nc as a single entity.

16.2.5.6 STS Frame Payloads - ATM Cells

During the design and standardization process for ATM and the SONET frame, specific steps were taken to support transmission of ATM protocols over SONET efficiently. When ATM protocols are carried by SONET, ATM cells occupy the payload capacity in the SONET frame. The ATM layer is mapped into the payload portion of the SONET frame, along with an ATM header, which assists ATM to maintain the overall end-to-end integrity of the user application. See Figure 16-4.

In the example of transporting a fixed alignment large payload of 149.76 Mbps (an ATM cell), a STS-Nc configuration is formed by linking N constituent STS-1s together in fixed-phase alignment. The SONET SPE carrying an STS-3c contains 261 (3x87) byte columns by 9 rows. The first column contains only one set of STS Path Overhead (POH) for an STS-3c SPE. Thus, through the Super Rate arrangement, STS payloads can be transported at higher rates and, where alignment is required, with low overhead.

When using SONET as an ATM transport system, instead of using the ATM Protocol Reference Model, the Physical layer can be presented as SONET layers (see Figure 13-2, page 100). This layer presentation supports the seamless interworking signal hierarchy and assists a carrier in maintaining the contracted QoS for ATM over SONET. The SONET layers (Path, Line, and Section) are discussed further in Section 17.

ATM cells (which consist of a 5-byte cell header and a 48-byte payload) can be mapped into SONET frames at a STS-3c or STS-12c Synchronous Payload Envelope. As mentioned earlier, the STS payload capacity is not an integer multiple of the ATM cell length, so the ATM payload will cross envelope boundaries. SONET requirements specify that ATM cells map to the STS-3c (260-column payload capacity) and STS-12c (1040 column payload capacity) SPE in such a way as to fill the entire payload capacity of the STS frames. The resulting

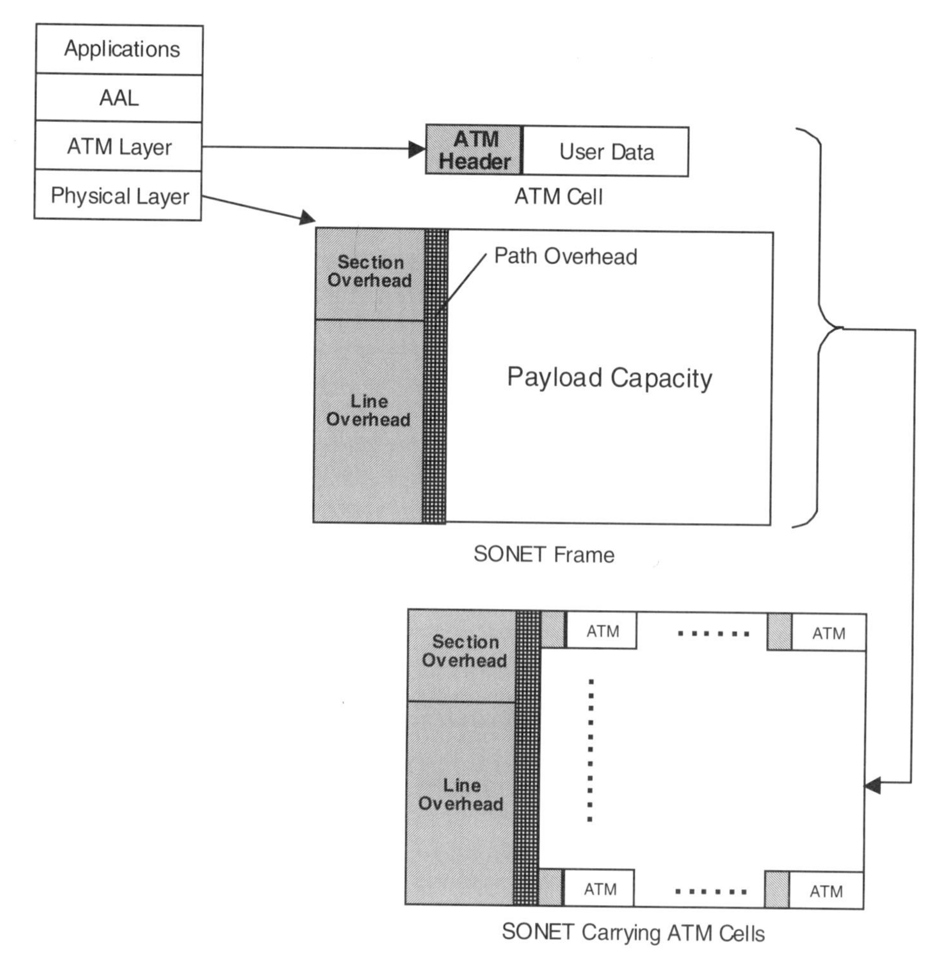

SONET Carring ATM Cells
Fig. 16-4

transfer capacity for ATM cells follow:

- ATM cells mapped to a STS-3c payload capacity yields a transfer capacity of 149.760 Mbps.
- ATM cells mapped to a STS-12c payload capacity yields a transfer capacity of 599.040 Mbps.

16.2.6 STS Frame Payloads - Virtual Tributaries

STS Payload sizes often vary from the capacity of one full STS Payload Envelope. This section will discuss a device generally used to handle smaller payloads.

16.2.6.1 Virtual Tributaries - Smaller Payloads

In telephony or data networks, the problem arises of variable speed network elements or functions (such as: sub-STS-1, or sub-DS3 rates) that must to be meshed with the STS-1 high-speed or backbone network components. SONET must accommodate signals that are smaller than exactly the STS-1 size, for example a sub-rate asynchronous digital hierarchy signal such as DS1. This enables SONET to connect with most existing transmission networks and sub-networks mapping into one standard transport mechanism.

To accomplish this, SONET must smoothly and efficiently integrate many lower speed connections into a higher speed SONET connection. The *Virtual Tributary* (VT) provides this function. The VT maps a low-speed digital signal into the STS-1 Synchronous Payload Envelope before it is converted into an optical signal. The VT structure accommodates sub-STS-1 payloads for transport and switching. Thus, a VT can carry a signal such as a DS1 or DS-2 within a byte-interleaved frame. VTs come in four sizes, see Table 16-1. It should be noted that DS3 is mapped directly into the STS-1 SPE and no VT structure is needed.

Virtual Tributaries are all constructed following the same basic concept. To explain this process, the DS1 map into the VT1.5 structure is described here. The DS1 payload contains 24 channels of information at 1.544 Mbps. This is considered a low-speed payload. The DS1 is mapped into the VT1.5, which has enough bandwidth to carry the DS1 payload and some path overhead. As shown in the first row of Table 16-1, the transport rate of the VT1.5 is higher than that of the original DS1 to account for the path overhead.

In this example, the DS1 plus overhead is mapped into 9 rows of 3 columns using a total of 27 bytes. This is the VT1.5 and its resulting capacity is:

$$\frac{27\text{ bytes}}{\text{frame}} * \frac{8\text{ bits}}{\text{byte}} * \frac{8000\text{ frames}}{\text{second}} = 1.728\text{ Mbps}$$

Bit stuffing is also required for the sub-DS3 signals to compensate for the difference between the bandwidth available in the VT SPE and that required for the actual asynchronous payload. Virtual Tributaries are used for mapping DS1C and DS2 signals. These signals are constructed in the same manner as DS1, however the VTs contain a larger number of columns to accommodate the larger signal rates[15].

The VT construction process is different than the STS-N signal built from

Signal Name	Payload Capacity (Mbps)	VT Size	Rate (Mbps)	STS-1 SPE Columns
DS1	1.544	VT1.5	1.728	3
E1	2.048	VT2	2.304	4
DS1C	3.152	VT3	3.456	6
DS2	6.312	VT6	6.912	12

SONET Virtual Tributary Sizes
Table 16-1

[15] As can occur with any conversion in the network, errors can be introduced. Consequiently, additional error management may be needed. Byte interleaving provides an additional error management function.

STS-1s. In the case of building STS-N signals, no additional overhead is added. The STS-N bit rate, which is of a simpler structure, is *n* times that of the STS-1.

Virtual Tributaries are not mapped directly into the STS-1 SPE. Instead, they first are mapped into *Virtual Tributary Groups* (VT Groups). The VT Group architecture accommodates different types of VTs in a single VT-structured STS-1 SPE. VT Groups do not have any associated overhead or pointers.

VTs within a VT Group are interleaved by column into the SPE. See Figure 16-5. This interleaving uses 84 columns in the SPE (12 columns per VT Group x 7 VT Groups). The 12 columns of a single VT Group are not consecutive within the SPE. There are three additional columns. One is used for the Path Overhead (column 1 of the SPE). The other two are fixed stuff (columns 30 and 59). For example, the first VT Group of the SPE would be found in SPE columns 2, 9, 16, 23, 31, 38, 45, 60, 67, 74 and 81. Columns 30 and 59 are skipped.

To support efficient transport, Virtual Tributaries have a fixed structure and size. A VT Group has a fixed size of 12 columns within the STS-1 frame. As the reader will recall, there are 87 columns in the SPE for the STS-1. Thus in the STS-1 frame, a SPE to be arranged for the VT structure is divided into seven VT Groups. VTs (one or more) are placed into a VT Group. A given VT Group can contain only one size of VT, as shown in Figure 16-6. A VT Group can contain:

- Four 1.544 Mbps DS1 frames, or
- Three 2.048 Mbps E1 frames, or
- Two 3.152 Mbps DS1C frames, or
- One 6.312 Mbps DS2 frame

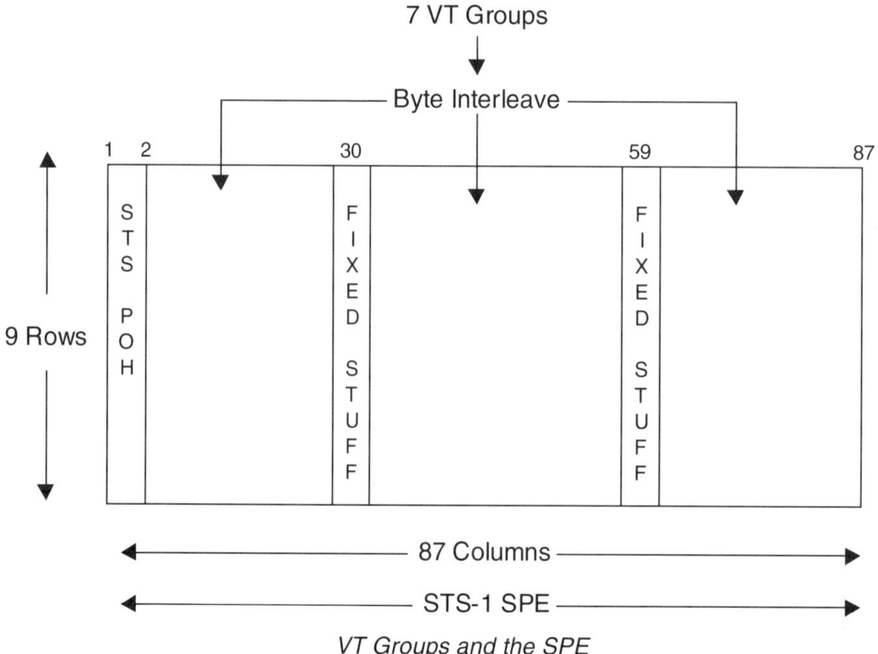

VT Groups and the SPE
Fig. 16-5

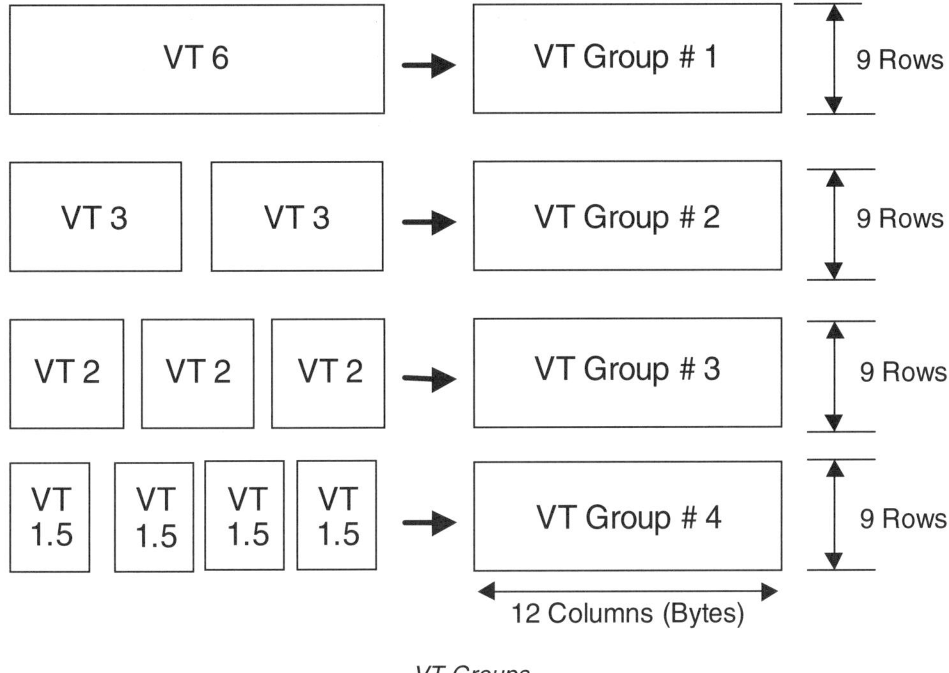

VT Groups
Fig. 16-6

A different VT size is allowed for each VT Group in an STS-1 SPE. Thus, an STS-1 SPE can contain a mix of VT sizes as shown in Figure 16-7. This flexible arrangement can accommodate a wide range of applications.

To understand a VT implementation recall Figure 14-5 (page 114). SONET ADMs in the terminal configuration could be deployed at the Central Office (the near-end, in this example) and connected to a Remote Digital Terminal (RDT) near the end of the loop plant (the far-end, in this example). DS1 signals are transmitted from the Central office to the near-end Terminal configuration ADM through a Central Office Terminal (COT). At the far-end ADM, another Terminal configuration ADM sends outgoing DS1 signals to the RDT for final termination in the network. In this example, all locally switched and non-locally switched traffic is grouped into a VT1.5 for each DS1. DS0 level grooming is performed at the termination point in the network (the RDT or COT). Further grooming (or facility fill), if needed, is performed in the ADM or the DCS before the signal reaches the local switch.

VT Operations Configurations

The SONET VT has two Operations configurations: floating and locked.

The *floating configuration* of VT uses a flexible mapping method to adjust locations of VT Groups within an STS-1 SPE payload. Pointers show the payload position in the STS frame. VT pointers enable easy access to synchronous payloads without the use of 125 μs buffers between the VTs.

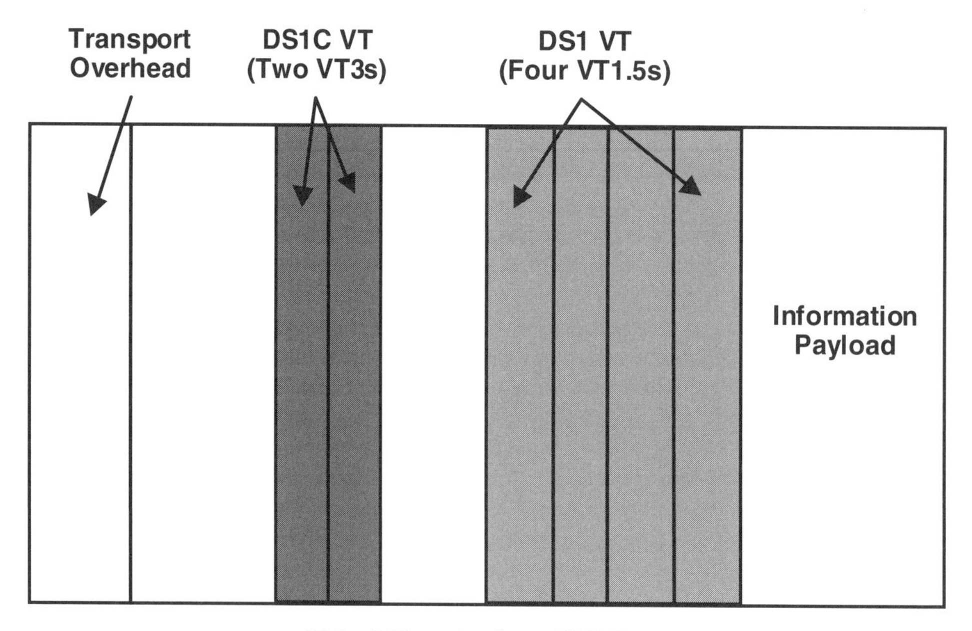

Virtual Tributaries in an STS Frame
Fig. 16-7

The *locked configuration* uses a fixed mapping method to identify VT Groups within an STS-1 SPE payload.

Both configurations allow different sizes of VT within an STS-1 SPE.

VT Mapping - Superframe Structure

The VT pointer is used also in the Virtual Tributary structure called the VT Superframe.

In addition to the division of VTs into VT Groups, a larger 500 μs structure (four consecutive STS-1 frames of 125 μs for each frame) is available. This larger structure is called a *VT Superframe*. Specific overhead bytes are established for a Superframe structure. For example, the H4 byte, in the STS Payload Overhead, indicates the phase of the pointers in the VT Superframe.

The first byte in each of the four VT payload frames in the Superframe is a pointer. The V1, V2, V3 and V4 bytes are payload pointers within the VT path overhead that indicate the address or position of information within a Superframe. These Payload Pointer bytes V1 to V4 are referred to as the VX byte, where the "X" is 1, 2, 3 or 4, depending on which VT frame in the Superframe is discussed.

After the overhead pointer byte in the payload is allocated, 26 bytes remain in the VT. These 26 bytes are called the *VT Envelope* which carries the *VT SPE*. This structure is analogous to that of the STS frame and STS SPE. Here, the VT payload pointer allows flexible and dynamic alignment of the VT SPE within the VT Superframe, independent of other VT SPEs. The use of this type of pointer minimizes the need for slip buffers. Slipping is the repeat or deletion of a frame of information to correct frequency differences. Slip buffers provide the capacity to perform the repeat or deletion function.

Thus, the capacity of the VT SPE in the Superframe structure (for example, for DS1) is:

$$26\ \frac{\text{bytes}}{\text{frame}} * 8\ \frac{\text{bits}}{\text{byte}} * 8000\ \frac{\text{frames}}{\text{second}} = 1.664\ \text{Mbps}$$

This is slightly lower than the VT capacity of 1.728 Mbps, listed in Table 16.1, because this capacity excludes the one pointer byte (VX) for each frame.

16.3 SONET PAYLOAD SIGNAL MAPPING FLOW AND DIGITAL HIERARCHY - SUMMARY

SONET provides flexible mapping and transport of many types of existing services and the transport foundation for many new services. The basic payload mapping flow of SONET is summarized in Figure 16-8, which shows the flow of sub-rate and DS3 signals into the SONET signal structure for optical transport (OC-1 and multiples thereof). This Figure also shows the mapping of STS-Nc for payloads greater than the STS-1 rate, such as ATM, as well as the existing asynchronous digital transmission hierarchy consisting of the DS1, DS1C, DS2 and DS2 signals.

To tie in the digital hierarchy referred to in Section 13 and Table 13-1 (page 103), look at the North American Digital Hierarchy with SONET signals, bit rates and payload relationships in Table 16-2. One of SONET's attributes is that it can transport many different digital signals using a standard Synchronous Transport Signal (STS) format.

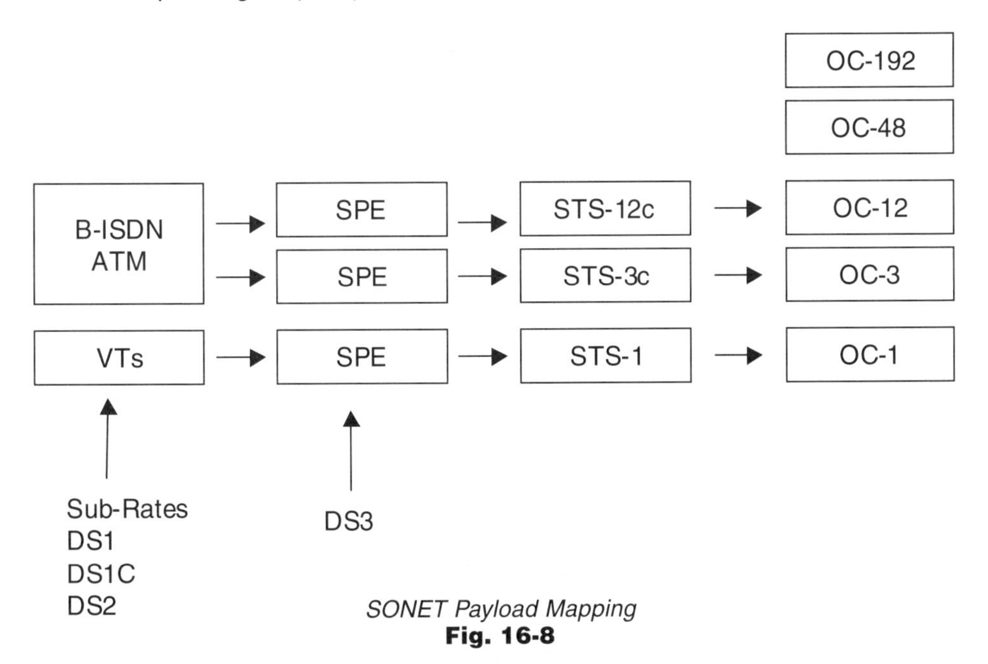

SONET Payload Mapping
Fig. 16-8

The payload capacity of SONET levels often is expressed in terms of DS1s and DS3s because these represent the dominant payload applications. The OC-N/STS-N signals can be a combination of payload types as long as the total capacity or bandwidth of the signal is not exceeded.

For example, the OC-12 can transport 9 DS3s and 84 DS1s. This can be expressed as SONET's narrowband capacity for OC-12, where "narrowband" is defined as a rate of DS3 and below. The equivalent DS3 capacity is equal to the *N* in the STS-N/OC-N designation. As an example, OC-48 has a DS3 equivalent capacity of 48 DS3s. As an another example, *N* DS2 circuits can be mapped into an STS-N frame. Also, it can be said that OC-N can establish *N* DS3 circuits, or *N*x28 DS1 circuits.

Signal Level / SONET Reference	Payload Capacity	Payloads
DS0	64 Kbps	1 Voice Channel or Data Circuit
DS1/VT1.5	1.544 Mbps	24 DS0s, or 1 DS1
DS1C/VT3	3.152 Mbps	48 DS0s, or 1 DS1C
DS2/VT6	6.312 Mbps	4 DS1s
DS3	44.736 Mbps	28 DS1s
/STS-1/OC-1/EC-1	51.849 Mbps	1 STS-1 (such as 44.736 Mbps DS3)
/STS-3/OC	155.520 Mbps	3 STS-1s or 1 STS-3c
/STS-12/OC-12	622.080 Mbps	12 STS-1s or 4 STS-3c
/STS-48/OC-48	2.488 Gbps	48 STS-1s or 16 STS-3-c or 4 STS-12c
/STS-3c/OC-3c	155.520 Mbps	≈ 50 Mbps to ≈ 150 Mbps (ATM)
/STS-12c / OC-12c	622.080 Mbps	≈ 150 Mbps to ≈ 600 Mbps (ATM)

VT = Virtual Tributary

STS = Synchronous Transport Signal

OC = Optical Carrier (a SONET Optical Signal)

EC= Electrical Carrier (a SONET Electrical Signal)

Concatenated = Continuous pipe / linked together (e.g. STS-3c)

North American Digital Hierarchy, Including SONET [16]
Table 16-2

[16] VT2 is a SONET Virtual Tributary, operating at a Payload Capability of 2.048 Mbps, however this is used for E1 transport, which is common internationally and not commonly used in North America and is not included in the North American summary of speeds and capacities.

DS0, DS1 and DS3 payloads are transported on SONET commonly. However, the vehicles to do this in SONET are STS-1 and VT frames. SONET overhead is added to the payload signals and accounts for the difference between frame rates and payloads.

OC-24 is no longer used as a standard rate in SONET systems.

17 SONET LAYERS AND OPERATIONS

17.1 SONET LAYERS

The layers of SONET (or operations flows at the SONET layer) are established to support efficient transport, synchronization, multiplexing, maintenance, provisioning and monitoring of the SONET system. These layers are the next building block in the SONET system. They tie together the SONET Network Elements and Frame structure in a logical arrangement, which helps integrate the SONET platform for overall system operation. These layers provide information about the transport only when and where it is needed. This process balances the need for operational information with the capacity and transport time necessary to transmit and process that information. This supports the overall efficiency and high reliability of the SONET system.

The layers of SONET are the Path layer, Line layer and Section layer. These layers contain operational information critical to SONET.

The SONET frame overhead structure and functionality are based on these layers. The operational information in these three layers is used to report SONET-level alarm, fault, status and error indications. It also forwards error detection information for the detection of transmission errors and collects SONET-level performance data. This section will overview the functions of these layers and how they work together to support the service provider's operations information needs.

Each SONET layer requires the services of all lower-level layers to perform its function, see Figure 17-1. For example, when two Path layer processes exchange DS3s, the Path layer maps the DS3 signal and the STS Path Overhead into an STS-1 Synchronous Payload Envelope (SPE). The SPE is given to the Line layer as an internal Path layer signal. The Line layer multiplexes several SPEs from the Path layer and adds Line overhead. During this process, the Line layer frame and frequency aligns each SPE from the Path layer. Then, the Section layer adds Section overhead and performs scrambling before transmission by the Physical layer, where the actual OC-N signal is formed, as the electrical signal is converted to the optical format.

Network Elements process the SONET Management Layers to transport and terminate frames and to support

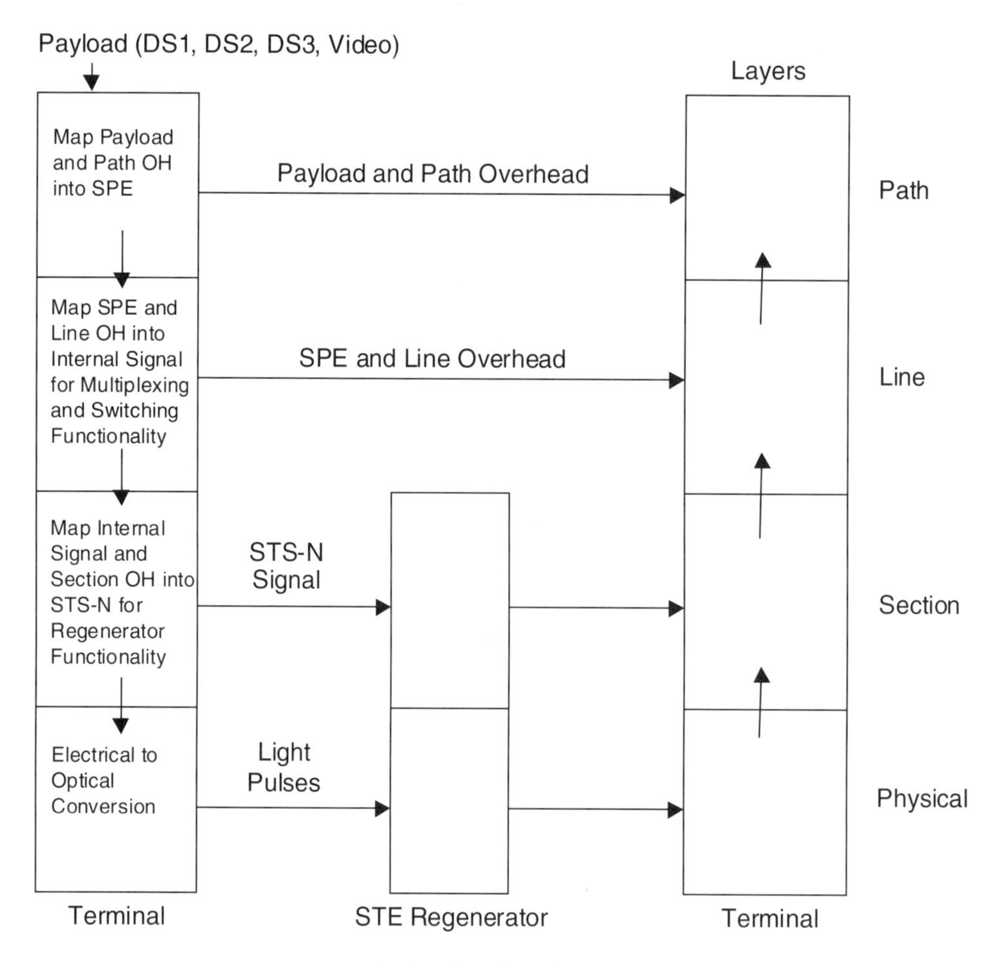

Optical Interface Layers
Fig. 17-1

SONET management systems. For example, the terminal NE terminates all three of these layers and, thus, is required to process the three corresponding signal fields. An end-to-end connection in SONET is a *Path* between two pieces of Path Termination Equipment (PTE). A Path is broken into segments called *Lines* between two pieces of Line Termination Equipment (LTE). The Line is further broken into segments called *Sections* which fall between two pieces of Section Termination Equipment (STE). See Figure 17-2.

Broad descriptions of the SONET Management Layers and their functions follow, along with examples of the Network Elements that process these Layers.

17.1.1 SONET Layers - Physical layer

The *Physical layer* supports the transport of bits as optical or electrical pulses across the physical medium. No overhead is associated with the Physical layer, nor is any operational information encoded at this level. The Physical

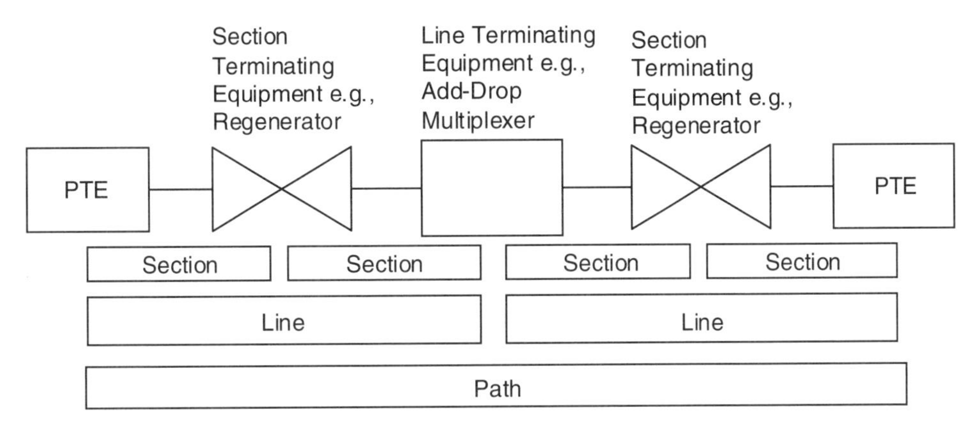

Path, Line and Section in SONET
Fig. 17-2

layer functions to convert between the internal STS-N signals and external optical or electrical SONET signals. This layer maps the various services into the STS-1 payload capacity for the Line layer. It also provides end-to-end performance monitoring, statistical reporting and Digital Signal (DS) to Optical Carrier (OC) mapping. The Physical layer adjusts pulse shapes and signal power levels. Equipment that operates at this layer includes electro-optical units, which convert between electrical and optical signals.

17.1.2 SONET Layers - Section Layer

The *Section layer* provides the transmission of multiplexed STS frames (STS-N) across the physical medium and uses the Physical layer for transport. The Section overhead, used in this layer, relates to the regenerator Sections of SONET. Network Elements that process Section layer information include the ADM in both the Add-Drop and Terminal configurations. This layer performs functions such as framing, scrambling, Section error monitoring and Section level communications (including the Local Orderwire or LOW used for local provisioning and maintenance functions). The Section overhead is interpreted and modified or created by Section Terminating Equipment (for example, one of the functions of a Terminal ADM is to terminate the Section Layer).

The Section and Physical layers can be used in some equipment including certain STS regenerators without involving the higher layers.

17.1.3 SONET Layers - Line Layer

The *Line layer* provides for the reliable transmission of the Path layer payloads across the physical medium as well as synchronization and multiplexing of STS-1 channels. All lower layers exist to provide transport for this layer.

Line layer overhead (in the Transport Overhead portion of the STS frame) also provides the overhead used for the maintenance span and line protection (between protection switches). Line overhead is interpreted and modified/ created by the Line Terminating Equipment. To access the Line overhead, the Section overhead must be terminated. Thus, a NE that is Line Terminating

Equipment also is Section Terminating Equipment.

An example of system equipment that communicates at the Line layer is an OC-M to OC-N multiplexer or an ADM in the Terminal configuration.

17.1.4 SONET Layers - Path Layer

The *Path layer* provides for the transport of various payloads between the SONET terminal multiplexing equipment. Examples of such payloads are DS1s and DS3s (see Figure 17-1 above). The Path layer maps to payloads into the format required by the Line layer. This layer communicates end-to-end by the Path Overhead (POH). The POH is interpreted and modified or created by Path Terminating Equipment (PTE). To access the Path Overhead, the Section and Line Overhead first must be terminated. Thus, Path Terminating Equipment (a Network Element which originates or terminates end-to-end user signals such as an STS-1 path) also is Section and Line Terminating equipment.

An example of the system equipment that communicates at this level is a DS3 to STS-1 mapping circuit.

17.1.5 SONET Layers - Layer Interaction

The SONET has a specified means of interaction for its layers. This basic structure is another of the building blocks that support SONET's flexibility and efficiency. As shown in Figure 17-1, each layer communicates horizontally to peer equipment in that layer. Also each layer processes certain information and passes it up to the adjacent layers, thus horizontal and vertical communications occur.

For example, in Figure 17-1, payloads are input to the Path layer. This layer transmits the payloads and the Path Overhead - POH horizontally to its peer entities. The Path layer maps the payloads and POH into SPEs. The SPEs are then passed vertically to the Line layer as internal Path layer signals.

The Line layer transmits these SPEs to its peer entities along with Line overhead. This layer maps the SPEs and Line Overhead into internal Line layer signals. The SPEs are synchronized and multiplexed and then the Line layer signals are passed to the Section Layer.

The Section layer transmits STS-N signals to its peer entities. This layer maps the internal Line layer signals and the Section overhead into an internal STS-N signal that is handed to the Physical layer for transmission. The Physical layer transmits optical or electrical pulses to its peer entities.

Not every SONET NE requires access to all layers. For example, a STE regenerator would use only the first two layers (Physical and Section). In the same fashion, a Network Element that only routes SPEs and does not drop or add any new circuits would use only the Physical, Section and Line layers. Only when a circuit is to be dropped or added from the NEs are all the layers accessed.

17.2 SONET LAYERS - OTHER ASPECTS OF LAYERS AND MANAGEMENT - LAYER TERMINATING EQUIPMENT

Each of the logical layers of SONET can be terminated in an appropriately configured Network Element.

Path Terminating Equipment (PTE) is a type of Network Element which originates or terminates the SONET Path layer. The PTE assembles and disassembles the standard frame format. PTE terminates the Path layer, then accesses, interprets and modifies the path overhead. For example, ADMs or DCSs are PTE.

Line Terminating Equipment (LTE) can be any Network Element that operates at the Line layer, such as an ADM or DCS. A LTE Network element cannot be a regenerator, because the Line layer is not available to the regenerator. The LTE terminates the Line layer, then accesses, interprets and modifies the line overhead.

Section Terminating Equipment (STE) can be any Network Element including regenerators. The STE terminates the Section layer, then accesses, interprets and modifies the section overhead.

17.2.1 Network Management - Performance Monitoring

SONET is designed to include bit oriented operational information encoded in the Section, Line and Path overhead fields. These bits are used to report SONET-level management information such as alarm, fault, status and error indications An example of such a condition indicator is the Far End Bit Error (FEBE) indication. These bits also provide forward error detection information for the detection of transmission errors and the collection of SONET-level performance data. All flows are bi-directional. Performance Monitoring processes are discussed in more detail later in this section.

17.2.2 Network Management - Data Communications Channels

To support its extensive network management functions, SONET requires accessible communications capability. It must link all significant SONET NEs and network monitoring points. SONET has two data communications channels (DCCs) to support this need. These channels occur within the optical carrier OC-N signal:

1. One channel monitors communications between terminals and repeaters. This channel is designated Section DCC (192 Kbps).
2. One channel monitors communications between NEs. This channel is designated Line DCC (576 Kbps).

The DCCs use message-based information transfer to communicate command, control and alarm information. Both the Section and Line DCCs support the same protocol and are connectionless and transaction-oriented.

The DCC establishes communications between the NEs. As a result, remote supervision or provisioning can be established in SONET. Other applications of the DCC are discussed further in Section 17.3.

17.3 SONET OPERATIONS

17.3.1 Operations Systems - Overview

One of the important aspects of SONET is the capability to support and interface easily to Operations Systems (OSs) for maintenance, performance monitoring and provisioning. OSs, themselves, are external to SONET Network Elements, but some functions are provided by the NEs themselves. OSs for SONET are continually evolving. This section will address functions that support SONET network management [38].

17.3.2 OS Types and Structure

Operations Systems can take many forms. OSs can be large capacity computer systems with many complex subprograms, smart computer terminals or other portions of the network, such as terminals, controllers and gateway elements.

An OS is any system that can send and retrieve messages from SONET Network Elements and act on those messages to perform the OS functions, such as monitoring, alarm receipt, provisioning, supervision and remote maintenance.

SONET overhead structures provide the capability to monitor the health and maintenance of payloads. For example, the STS-1 Section overhead contains the B1 byte. This byte is a parity check. At the receiving end of SONET, the B1 byte is compared against a newly calculated parity check. Error indications are generated by performance monitoring systems if there are discrepancies between these two parity checks.

OSs uses the Data Communications Channels that exist within SONET Section and Line overheads. Through DCCs, OSs can access SONET Network Elements (NEs) to monitor, supervise and provision.

Computer terminals (and other sub-network elements) are required to place messages to, or retrieve messages from, the Network Elements. Using the DCC Section and Line overheads, a computer terminal can remotely supervise and provision the NEs in SONET.

The following sections review how SONET and the OSs communicate.

17.3.2.1 OS Communications

SONET operations communications can have different topologies depending on the configuration (intra- or inter-site) and application (OS-NE, NE-NE, survivable rings). However, all SONET operational communications are based on a common set of OSI protocols collectively referred to as the SONET Operations Communications Interface.

The following sections will discuss NE-to-NE, NE-to-Maintenance Technician and OS-to-NE operations communications.

NE-to-NE Communications and Provisioning

SONET NEs need a communications method to exchange alarm, failure, status and error indications. This communications can occur through Embedded Operations Channels (EOCs) or LAN connections.

EOC communications, through the Section DCC, can be message or bit oriented. For example, if bit 5 of the G1 byte is set and conveyed, the Remote Detection Indication (RDI-P) alarm is transmitted. If the B1 byte is set (8 bits), the Section alarm Bit-Interleaved Parity (BIP-8) is transmitted. Once these bits (or bytes) are received and processed, the NE can take appropriate action to support high levels of network performance.

NE-to-NE communications also could be established through an external LAN. For example, an Ethernet LAN could provide download software changes and system upgrades to SONET NEs. This arrangement can provide another powerful dimension to the overall efficiency and effectiveness of a SONET system.

Ethernet can operate at the 10 Mbps rate over coaxial cable, twisted pairs or fiber optic cable; however, in some cases special equipment and cabling network may be needed to support the LAN for this type of NE communications. Ethernet IEEE 802.2 and 802.3 specifies LAN implementation in SONET standards.

NE-to-Maintenance Technician Communications

Effective and efficient communications between the network Maintenance Technician (MT) and NE are important. The MT can communicate directly with the NE for network monitoring and maintenance functions using terminals, laptop computers or workstations.

The current protocol used to communicate between the MT and NE on the Digital Communications Channel is the Transaction Language 1 (TL1). An upgrade to this Language is under development and is called the Common Management Information Service Element (CMISE). Additional enhancements have been provided by some SONET NE manufacturers to reduce the number of keystrokes for certain functions and to reduce the memory requirements on terminals and interactive devices. These enhancements can support storage and retrieval of groups of commands to reduce the time-to-service of installations, turn-ups and component/system upgrades.

OS-to-NE Communications and Provisioning

Communications between OSs and NE can be direct or indirect. A direct communication requires a single OS-NE interface. Indirect communications typically involve at least one NE-to-NE interface and/or a gateway function.

Network elements have varying access to the OS DCCs depending on their configuration type. See Table 17-1.

OS-to-NE communications could also be established through an external LAN, such as the Ethernet LAN (example described previously).

Mediation Device-to-NE

A *Mediation Device* (MD) can perform some gateway functions. The primary purpose of a MD in a subnetwork is to facilitate the OS-to-NE communications. It also can support access from a remote Maintenance Technician.

NE Type	Connection	Communication Functions Performed
Gateway Network Element	Connects Subnetworks to OSs	Routes Traffic, Converts Protocols, Maps Addresses, Converts Messages, Concentrates and Distributes Messages, Resolves Interworking Issues
Intermediate Network Element	Connects Subtending NEs	Routes and Relays Tandem Traffic on Same Network Type, Has Subtending NEs
End Network Element	Subnetwork Termination	Only Handles Own Traffic

NE Types for OS Communications
Table 17-1

The MD is not part of the SONET network but can a perform stand-alone, non-traffic functions, such as:

- Statistically concentrating packets for the OS-to-NE interfaces of many SONET NEs. This can reduce or eliminate individual dedicated links between each SONET NE and the OS.
- Filter alarms between a SONET gateway NE and the OS. For example, the MD could prioritize alarms and control the flow of a large and rapid sequence of alarms if a fiber or node failed.

Other OS Communications Considerations

Some manufacturers have designed proprietary management systems for their own subnetworks. These controllers can intensively monitor and supervise SONET networks. Because these management systems are proprietary, a mediation device must interface these subnetwork controllers with other supervisory elements of the overall SONET system.

OSs are not part of the traffic-bearing portion of SONET. Different manufacturers take different approaches to OS development, therefore compatibility issues may exist between OSs.

17.3.3 SONET Maintenance Overview

The major feature of SONET's maintenance capability is the number of features built into the system to support high quality and reliability through effective maintenance. This is one of the major factors that differentiates it from earlier high-speed transmission schemes. The maintenance features of SONET include alarm surveillance, performance monitoring, testing and control features for certain operations of NEs. This section will discuss some of the major elements of the maintenance features and processes of SONET.

Major maintenance tasks supported by SONET include:

- Trouble detection. It can detect and declare failures.
- Trouble or Repair Verification. It can verify the continued existence of a problem.
- Trouble Sectionalization. SONET can sectionalize the failure to a NE.
- Trouble Isolation. It can isolate the failure to a replaceable circuit pack, module or fiber.
- Restoration. SONET can restore service, possibly before the isolated failure has been repaired.

To accomplish these functions, SONET must methodically identify, isolate and test faults, and (possibly) reroute traffic to restore service. The NEs play a key role by generating appropriate messages, transmitting messages, performing tests/loopbacks and diagnostics. Certain NEs can reroute traffic around trouble spots (for example, ring topology protection or the DCS) and assist in restoring service. The SONET maintenance capabilities start with an organized and consistent approach to identifying and labeling problems. These features are designed into the system.

17.3.4 Performance Monitoring

In a comprehensive maintenance program, which would exist with a typical SONET installation, failure detection is key. SONET NEs generate failure

Loss of Frame
Loss of Pointer
Equipment Failure (such as Fuse or Power, CPU, Optical Receiver or Transmitter)
Protection Switch Byte Failure
Channel Mismatch Failure
Far - End Protection Line Failure
STS Payload Label Mismatch
VT Payload Label Mismatch

SONET Network Element Defect Indicator Examples
Table 17-2

indications according to the layers of function provided by the NEs. One important strength of SONET is the establishment of a minimum set of failure detection indications for various NEs.

Failure detection indicators support an important feature of SONET called the *Performance Monitoring* (PM) capability. To be fully effective the PM capability must be established in a structure where the other facets of SONET standards have been applied. This could include SONET NEs, SONET topologies and standard frame structures, as well as an OS capable of working with SONET, appropriate gateways, a well established maintenance strategy and trained personnel. Examples of defect indicators are listed in Table 17-2.

Performance Monitoring is in-service, non-intrusive monitoring of transmission quality. SONET has established a structured system of communicating failure conditions as a part of PM. SONET accomplishes this type of monitoring by accumulating PM overhead bits including BIP-Ns in the Section, Line, STS Path and VT Path layers. PM information is available at the Physical layer and with DSns using physical parameters.

SONET accumulates PM overhead parameters in registers (such as current period, previous period, recent period and threshold). See Table 17-3. By using these registers, PM enables these parameters to be presented to the OS and/or Maintenance Technician as defect indictors for patterning or action.

PM overhead parameters can include:

- Optical Power Received
- Optical Power Transmitted
- Laser Bias Current
- Pointer Justifications
- Protection Switching Duration
- Protection Switching Counts
- Failure Counts
- Unavailable Seconds
- Severely Errored Seconds
- Errored Seconds
- Coding Violations

Registers
Current 15 Minute
Current Day
Previous 15 Minute
Previous Day
Recent 15 Minute

SONET Performance Monitoring Register Examples
Table 17-3

As an example of the Performance Monitoring process, PM parameters are accumulated in registers (e.g., See Table 17-3) and compared against thresholds. When the parameters reach the threshold, the information is presented to the OS and if necessary to the MT. If all parameters were presented to a Maintenance Technician with equal weight, the technician would be overwhelmed and scarce maintenance resources would be misdirected. Therefore, performance parameters must be accumulated in a buffer or register until a preset threshold has been reached. Then, an indicator is forwarded, for example, to a Operations System for appropriate action.

As Performance Monitoring indicators are forwarded to OSs and NEs, an automated response can occur. For example, the system could automatically reroute or reconfigure traffic. The network operator can establish preset responses (such as automatic rerouting to protection pairs on a ring) to these indicators.

The OS analyzes and further refines the register thresholds and patterns and a Maintenance Technician can be alerted for intervention or on-site maintenance, as needed. The alerts are filtered by the OS to limit the quantity presented to the MT. Only those of sufficient priority are presented to the MT for intervention. Following the receipt of an indicator, a MT could initiate a test or dispatch a field repair technician based on the threshold message. The priority and weighting of alerts sent to the MT is set by the network carrier or operator.

17.3.4.1 Alarms

All SONET NEs are required to detect certain defects on the incoming signals relevant to the layers of function that a NE provides. A defect could be a limited interruption or a larger failure. The detection of a defect could cause an action, such as the transmission of a maintenance signal. However, when the defect condition persists, the NE sets a failure indication. This indication may or may not be sent to the OS, depending on the level of severity and the thresholds for trouble conditions designed into the particular system.

The next section focuses on alarm hierarchy and processes. Alarms are generated when defects occur in a persistent enough manner.

SONET Alarm Hierarchy

A SONET maintenance alarm signal (or Alarm Indication Signal-AIS) uses a hierarchy (see Table 17-4) to quickly direct maintenance efforts to the appropriate portion of the network and to signify the level of alarm. The higher the level of

Alarm Level	Alarm Example	SONET Layer
Highest	AIS - L	Section to Line
	AIS - P	Line to STS Path
	AIS - V	STS Path to VT Path
	DS0 AIS	VT Path to DS0 Path
Lowest	Trunk Conditioning	Below DS0 Path

SONET Alarm Hierarchy
Table 17-4

the alarm, the greater the magnitude of the alarm condition. Once the AIS is set, it is sent downstream to the next SONET NE.

The AIS-L message alerts the downstream Line Terminating Equipment that a defect has been identified on the incoming SONET Section. It may also indicate that the LTE supporting provisioned line origination functions failed. If Line-level Automatic Protection Switching is provided, the downstream NE could initiate protection switching upon receipt of the AIS-L signal.

If SONET is the transport medium for lower-speed digital signals (such as DS1 or DS3), NEs could generate a DSn AIS. A DS0 AIS may not be applicable outside the SONET network. Therefore, an interface to a non-SONET NE may require the application of a service-specific trunk conditioning code.

An indication to upstream terminals can be sent in a SONET system by the expanded use of Remote Alarm Indication (RAI). Remote Detection Indication (RDI) signals in SONET can occur at the Line, STS Path and a VT Path layers to support this function. A Remote Failure Indication (RFI) also is available to provide backward compatibility to equipment using traditional RAI signal timing and for translation to and from DS1 RAI signals.

Where simplex (one-way) payloads are multiplexed, some of the functions of SONET may not be applicable. This includes STS or VT RDI and REI generation by Path Terminating Equipment.

SONET Alarm Processing

SONET alarms are processed similar to conventional telephony alarm monitoring methods. They alert the system to the existence, location and severity of the trouble. Trouble notifications include output messages, which are visual indications at the NE. Also, trouble notifications display audible and visible indications on the frame or equipment bay that houses the NE. A trouble notification message would include:

- The Type of Trouble
- The Time Event Occurred
- The Signal Level Affected
- A Service Affecting (or Non-Affecting) Indicator
- An Alarmed (or Non-Alarmed) Indicator
- The Alarm Level Indicator (e.g., Critical, Major or Minor)

Some SONET-specific modifications have been made to this traditional method of alarm processing. For example, an NE using linear Automatic Protection Switching would designate a STS Loss of Pointer (LOP) message as service-affecting[17], whereas the OC-N Loss of Synchronization (LOS) would be considered non-service affecting if the traffic is restored by a successful protection switch.

Another example of SONET-specific modification is that traffic-related failures can be grouped into near-end (e.g., LOS, LOF, LOP) or far-end failures (RFI). This permits SONET to group multiple RFI failures because these failures do not disrupt the capability to detect defects. This simplifies the reporting of SONET system defects.

---

[17] Where a LOP is detected on an active line, the alarm indication is classified as service-affecting. Where the LOP is detected on a standby line, the alarm indication is classified as non-service affecting.

17.3.4.2 SONET Testing

Testing procedures are part of an overall maintenance program designed to isolate a failure and replace or repair the defective entity. Maintenance tools are built into the SONET signal format for Performance Monitoring. Additional SONET tools are test access, diagnostics and loopbacks. SONET also features specific control and restoration features for NEs.

The long-term goal for SONET testing is to evolve toward self-diagnosing NEs. The groundwork for this long-term goal is established in the Performance Monitoring and Alarm plans. Additionally, SONET diagnostic requirements provide the foundation for the evolution to self-diagnosing NEs.

SONET testing activities include:

- Analyzing Alarms, Performance Monitoring Data and Maintenance Signals
- Executing Diagnostics
- Executing Controls (Switch To Protection)
- Activating Loopbacks To Prepare For Testing
- Test Accessing For Signal Measurements

Operations personnel can gain access for these SONET activities through a local maintenance interface or the remote operations interface.

Test Access

In SONET, test access is arranged for non-intrusive monitoring and intrusive testing. SONET Network Elements are designed for three categories of test access:

- Access to the fiber for monitoring and testing the optical signals and the fiber.
- Access to the SONET signal for monitoring and testing the format, mapping and equipment specifications.
- Digital test access to lower-speed digital signals to test services. Note that VT programmable capabilities, which allow the flexible assignment of low-speed signals to timeslots within an OC-N signal, may also provide DS1 remote test access.

Diagnostics

Diagnostics are generally categorized in two ways; those internal to an NE and those external as applied to a user's or Maintenance Technician's testing tool.

Network Element manufacturers include diagnostics internally in the NE. NEs are manufactured to do some or all of these:

- Run continuously
- Run on a preprogrammed or user-defined schedule
- Trigger running based on an event

These internal diagnostics are most often designed to run during normal operation. The user is unaware of the operation of these diagnostics. However, some internal diagnostics are manually initiated (by the user) between automatically scheduled times. These diagnostics are referred to as "on-demand" diagnostics.

A Maintenance Technician's testing tool can be activated on demand to:

- Retrieve information from SONET overhead.
- Operate special circuitry to support trouble analysis, such as a corrupted Bit Interleaved Parity (BIP).

These tools are activated through a workstation or OS interface to support trouble analysis, such as to analyze a corrupted BIP or to run an NE diagnostic and analyze the result.

Typically, diagnostics can occur at the Physical layer, Section layer (involving a loopback internal to the SONET NE), Line layer and Path layer. STS and VT signals can be checked through signal label diagnostics.

Loopbacks

Two types of SONET loopbacks support pre-service operations and test-related activities. These are terminal and facility loopbacks. SONET NEs may need to provide loopbacks for both SONET and DSn signals to support some applications. A loopback test is a configuration where the path or circuit is broken and "turned around" to be sent back to the testing point. This configuration permits the Maintenance Technician, for example, to examine the terminal or NE at a distant location or examine the facility between the test point and the distant location. This examination can be done remotely using the loopback configuration without dispatching a MT to the distant location, saving considerable expense.

Typically, loopbacks interrupt the flow of traffic. Changing transmission direction requires coordination among the two or more NEs affected by the test. A loopback should not be used as a routine test, since the loopback test interrupts traffic and has a potential impact on the network.

In general, loopbacks are performed with the use of external test equipment to monitor the signal. SONET loopbacks can be performed with appropriately

Control Feature	Function
Reinitialize System	Hard Boot the System Reloads the NE's Operations System
Restart System	Soft Boot the System Reload Applications not the Operations System
Reestablish Failed Entity	Reconfiguration of Route Tables, or Reroute Facilities, etc.
Remove Entity from Service for Tests	Remove NE for Tests
Inhibit and Allow Indications	Suppress and Restart Messages from NE for Alarm and Non-Alarm Indications
Status Check	Check Equipment Configuration Status Including Active, Replicated Equipment and the Active Synchronization Source
Protection Switch	Switch Traffic to Protected (or Working) Fiber Facilities
Active Standby Switch	Manual Switch for Replicated Hardware or Software
Synchronization Source Switch	Manual Switch Between Sources

SONET NE Control Feature Examples
Table 17-5

equipped SONET terminals and facilities, as well as DSn terminals and facilities. However, exercise care when looping back a circuit to avoid unnecessary disruption of traffic flows.

Control and Restoration

Control features of the SONET NE are important in any carrier's performance and maintenance program. SONET NEs notify the OS when any control function has been executed. Examples of SONET NE control features and their application are shown in Table 17-5.

In summary, the features and functions designed into a SONET system permit a network or carrier operator to significantly reduce maintenance costs and improve system quality.

18 SONET SYNCHRONIZATION

■ 18.1 SYNCHRONOUS AND ASYNCHRONOUS NETWORKS

As a *synchronous* network, SONET requires effective and reliable synchronous design and operation of its Network Elements and protocols. This section will provide an overview of how synchronization is distributed in a network and the specific application of SONET synchronization.

An *asynchronous* network operates with its network elements controlled by independent clocking sources. In this type of network, matching transmit and receive pairs operate on approximately the same clock frequency. This means that the pairs are bit-synchronized and frame-synchronized. However, when the network looks across pairs, there is no provision for a common frequency of transmission.

A synchronous network, therefore, needs to synchronize its bits, frames and entire network. The NEs in a synchronous network are controlled by a common timing mechanism based on clocked signals. These elements are tied to a common time base, operate under the control of a common clocking source and are dependently timed.

Synchronous networks support more efficient access to signal low-speed interfaces (tributaries) and eliminate the need for back-to-back multiplexing. Synchronous networks also support a more efficient means of adding or dropping information from a digital data stream. A synchronous network does not require asynchronous bit stuffing, a process which is necessary in asynchronous networks where frequency variations from independently timed multiple inputs occur.

To support maintenance of synchronization, the SONET network uses pointers in the payload overhead structure.

■ 18.2 SONET SYNCHRONIZATION

Synchronization of the SONET network is based on a master-slave operation whereby a master clock distributes timing to other transmission systems (nodes) using the embedded transmission systems. The timing source (or master) feeds the system clocks the Network Elements. The control (or reference) of the timing in SONET Network Elements is traceable back to a master clock or Primary Reference Source (PRS). The system clocks in the NEs oversee most of the processes within the NE by providing the timing [39].

The synchronization or timing of SONET NEs can be established by ties to external, line or internal sources. The best timing source is an external source. An example of this type of source is a Building Integrated Timing Supply (BITS). Alternately, timing derived from the OC-N could be used. The least desirable alternative is an internal timing source.

18.2.1 External Timing Sources

SONET is responsible for the timing in a NE when external timing is used. The NE system clock can follow the frequency of the synchronous network via a synchronous link from that external network. External source synchronous links can supply both frequency and phase references to the NE system clocks. External timing is preferred due to administrative and maintenance savings.

When an external timing source is used, it is often referred to as a BITS clock. This clock is equipment dedicated to a single building or site for the timing needs of the equipment therein, see Figure 18-1. The clock with the highest accuracy within a building is designated as the BITS clock. All other clocks within the building receive timing from the BITS clock. This approach reduces dependencies on multiple clocks, in addition to reducing administrative and maintenance overhead and reducing transmitted errored seconds. Typically, all SONET NEs are externally timed from existing BITS clocks.

The overall telecommunications multi-carrier network is evolving towards more of a flat hierarchy whereby nodes receive timing from out-of-network sources, such as *Global Positioning System* (GPS), which is based on a system of satellites. Timing boundaries typically exist between countries or between two carrier's domains. In a typical installation, each domain would be synchronized and traceable to a primary reference source. This mode of operation is referred to as plesiochronous operation.

18.2.1.1 Stratum Timing Levels

External timing references are organized in a hierarchy known as Stratum

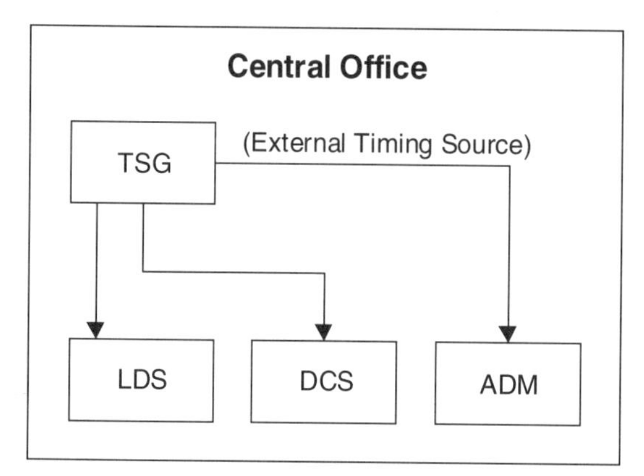

TSG - Timing Signal Generator
LDS - Local Digital Switch
DCS - SONET Digital Cross-connect System
ADM - SONET Add/Drop Multiplexer

Building Integrated Timing Supply - Block Diagram
Fig. 18-1

Minimum Accuracy	Stratum Level	Timing Reference or Network Element	Timing Source
1×10^{-11}	1	Primary Reference Source	GPS Clock
1.6×10^{-8}	2	Switch	DS1
4.6×10^{-6}	3E	DCS or ADM	DS1
4.6×10^{-6}	3	DCS or ADM	DS1
3.2×10^{-5}	4	Digital Loop Carrier	Composite Clock*

* Note: The 64 Kbps Composite Clock (CC) provides both frequency and phase references. DS1 commonly is used to supply a frequency reference in pre-SONET networks, however payload bearing DS1s should not be used to supply timing due to pointer processing that may occur within the SONET networks.

Stratum Level Synchronization
Table 18-1

Levels. These Stratum Level references normally are associated with the Building Integrated Timing Supply (BITS). Table 18-1 summarizes the accuracy requirements of the various Stratum Levels.

Figure 18-2 shows an example of Stratum clocks applied to SONET. NEs operate under the timing control of this master clock. The SONET NEs are controlled by an internal clock which is controlled by (or referenced to) a master clock. The master clock may have direct control over an NE. This means that the NE is tied to the BITS clock, which is referenced to the master clock directly or through a Stratum 2 reference. The master clock also could have indirect control of an NE. This means that the NE timing is tied to another NE clock which in turn is traceable back to the master or PRS. For example, the clock in the ADM in Figure 18-2 is controlled by the Stratum 2 clock which is slaved off the PRS. The Stratum 3 clock in the diagram can be traced back to the PRS through the SONET network to a Stratum 2 clock to the master. A BITS clock supplies timing to the elements in the building in which it resides (see Figure 18-1).

18.2.1.2 Other Timing Sources

If a BITS clock or other external timing source is not available, an alternative is to use OC-N derived timing, See Figure 18-2. A Network Element timing system can derive clocking information from an incoming OC-N signal directly from the bit transitions seen in the OC-N signal.

When clock recovery is attempted at a far end of the network and a reference source is needed, the OC-N signal is preferred to the traffic-bearing DS1 signal used in many networks. The advantages of OC-N derived timing follow:

- OC-N derived timing can be constantly available.
- OC-N derived timing can support Synchronization Distribution.
- Fiber optic cable, used for OC transport, is not affected by electromagnetic interference.
- The SONET topology, which uses OC-N facilities, has redundancy in the form of working and protection pairs.

Several timing choices exist when using OC-N timing. They are listed in Table 18-2.

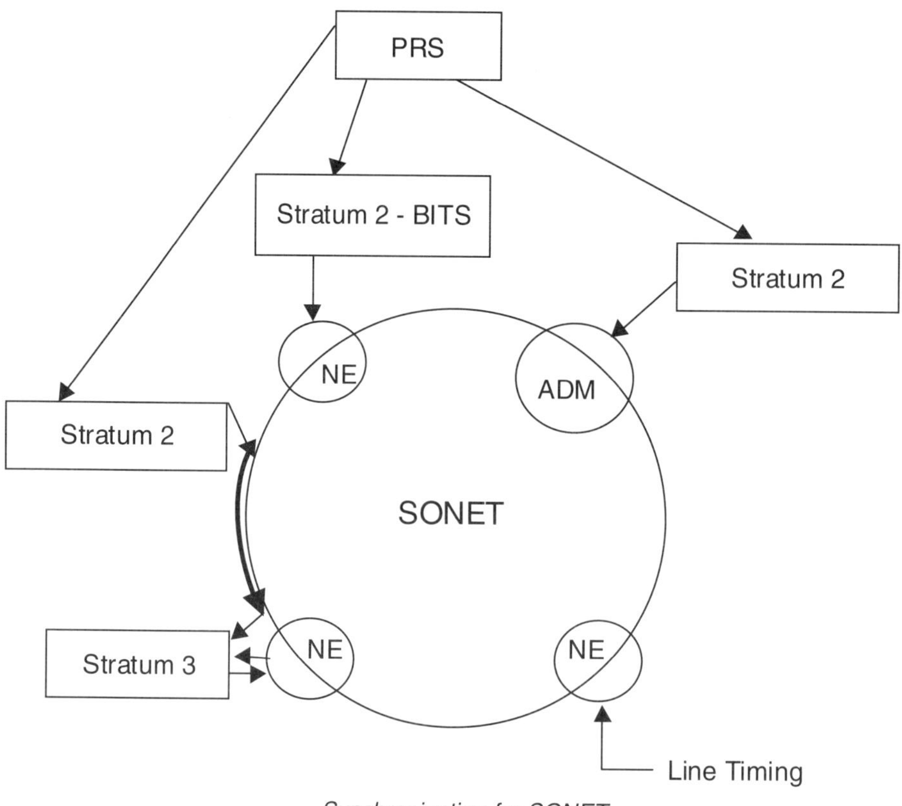

Synchronization for SONET
Fig. 18-2

An exception to timing SONET NEs from an external source (such as a BITS clock) is when NEs are located at the far end of the network where no BITS clocks exist. This could include, for example, terminal multiplexers in small end offices, the loop plant, certain customer locations or ADMs. Typically, these are line or through timed as appropriate.

Timing Type	Timing Derivation Method
Loop	NE supplies a single high speed output based on timing derived from an OC-N
Line	NE supplies both an "east and west" high-speed output based on timing derived from an OC-N
Through	NE supplies "east and west" output derived from separate timing signals from two OC-Ns

OC-N Timing
Table 18-2

To avoid using a reference signal from a payload-bearing DS1, a synthesized DS1 signal can be created within the SONET NE to establish a reference signal for the network clock. The synthesized signal can be based on the timing derived from OC-N timing. OC-N derived timing information can be passed via links between clocks in a synchronized network.

18.2.1.3 Internal Timing Sources

An alternative to SONET synchronization is to use the internal system clocks, found inside NEs. These clocks are less accurate than those tied to the synchronized network. For example, a NE can be equipped with a Stratum 3 internal clock, or a SONET Minimum Clock (SMC). These clocks have specifications that relate to free-run accuracy, entry into and restoration from holdover and holdover stability.

No type of "sub-SMC" clock has been standardized for use in applications where holdover may not be necessary (such as a line-timed SONET ADM in the terminal configuration that is transporting only bi-directional services).

19 OVERVIEW OF SIMILARITIES AND DIFFERENCES BETWEEN SONET AND SDH

Current standards and practices use SONET as the physical layer for transporting ATM cells. The international version of SONET, based on standards from the International Telecommunications Union (ITU), is the Synchronous Digital Hierarchy (SDH). The basic differences are minimal and highlighted in this section.

SONET has been developed and continues to be updated by the ANSI standards body, Committee T1 for North America. SDH has been developed in CCITT/ITU-T, the international standards body, over the same period of time. These two standards bodies work closely on the development of these two systems.

SONET transport starts at the basic rate of 51.84 Mbps and extends in multiples, thereof, to at least 9.95Gbps. SDH starts at the basic rate of 155.52 Mbps (or three times the basic rate of SONET). SDH extends in multiples to at least 9.95 Gbps, the same as SONET. See Table 19-1. The basic frame of SDH is called the Synchronous Transfer Module (STM)-1. STM-1 also refers to a signal in the optical format.

Most other differences between SONET and SDH are in nomenclature and in how the signal overhead is formatted, used or interpreted.

The basic frame (810 octets or bytes) in the SONET STS-1 consists of 90 bytes or columns transmitted in 125 μs and contains Transport and Payload Overheads. The basic frame in SDH is also transmitted on 125 μs and transmits an overhead and payload of 270 bytes (octets). However, the line rates are equivalent for corresponding SONET and SDH levels, as shown in Figure 19-1. SDH begins its basic trans-

SONET Level	SDH Level	Line Rate (Mbps)
STS-1/OC-1	-	51.84
STS-3/OC-3	STM-1	155.52
STS-12/OC-12	STM-4	622.08
STS-24/OC-24	-	1244.16
STS-48/OC-48	STM-16	2488.32
STS-192/OC-192	STM-64	9953.28

SONET and SDH Rates
Table 19-1

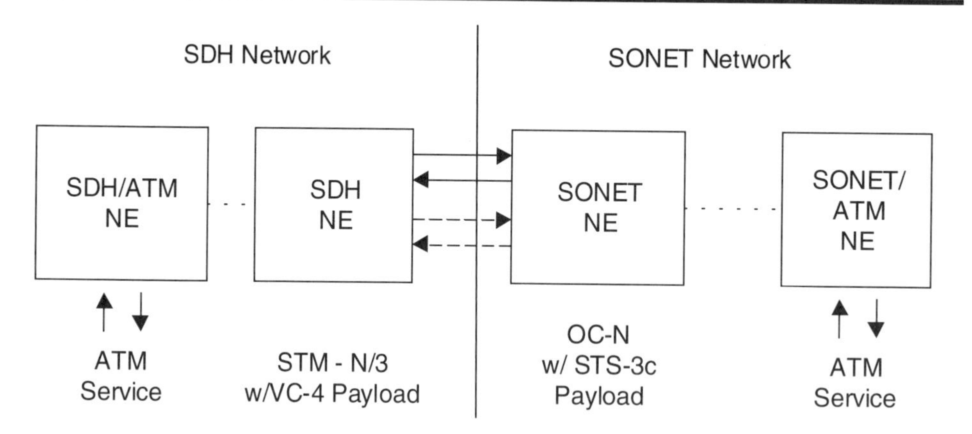

SONET/SDH Interface Network Example
Fig. 19-1

port rate at three times the STS-1 rate.

Higher-rate signals for SONET and SDH are formed in a similar manner of multiplexing and both SONET and SDH can carry a mixture of signals simultaneously.

In the transport overhead, both SONET and SDH have the same total number and layout of overhead octets at each transport layer. However, the first three rows of SONET transport are designated for Section Overhead. SDH refers to these as Regenerator Section Overhead. The first row of SONET OC-N Line Overhead is called Multiplex Section Overhead in SDH. In addition, some differences exist in the definition and/or usage of the overhead octets between SONET and SDH.

With regard to sub-rate signal mapping, SONET uses the Virtual Tributary arrangement described in Section 16.2. SDH uses a similar arrangement; however, sub-rate signals (E1 or E3, which is 34.368 Mbps), called Container (C) signals, are mapped to the synchronous transport signal through the use of Virtual Containers (VCs). VCs are composed of the payload signal and *path overhead.* The VC entity is transported transparently on SDH in much the same way as a VT is transported on SONET. SDH uses two types of VCs: *lower order* and *higher order.* VCs used in the SDH standard are shown in Table 19-2.

Lower-order VCs correspond to SONET VT SPEs. These are mapped to higher-order VCs through Tributary Units. These correspond to the VT in SONET. TUs are grouped into Tributary Unit Groups (TUG) in much the same way as VT Groups are for SONET.

Higher-order VCs correspond to SONET STS SPEs. For example, the VC-3 is similar to the SONET STS-1 SPE and the VC-4 is similar to the SONET STS-3c SPE.

Similar to SONET VT Groups, SDH establishes a fixed framework for TUGs with a Virtual Container plus a pointer. TUs are multiplexed into a TUG, where a TUG can hold only one type of container.

SDH permits different mappings of the same payload type, which differs

SDH Order	Signal Name	Payload Container	SDH Virtual Container	TU Capacity
Lower	DS1 (1.544 Mbps)	C-11	VC-11	1.728 Mbps
Lower	DS1 (1.544 Mbps) E1 (2.048 Mbps)	C-12	VC-12	2.304 Mbps
Lower	DS2 (6.312 Mbps)	C-2	VC-2	6.912 Mbps
Higher	E3 (34.368 Mbps) DS3 (44.736 Mbps) ATM (48.384 Mbps)	C-3	VC-3	48.960 Mbps
Higher	E4 (139.264 Mbps) ATM (149.760 Mbps)	C-4	VC-4	150.336 Mbps

SDH Containers
Table 19-2

from SONET. In addition, SDH requires four hierarchical levels for multiplexing: C signal to VC to TU, and the TUG to the Administrative Unit (AU) and to the STM. SONET requires three levels for multiplexing: sub-rate signal to the VT, to the VT Group, and finally to the STS-1.

SONET has a three-layer hierarchy for signal management (Section, Line and Path). SDH has four layers and pulls out the pointers from the frame overhead. The SDH management layers are: Regenerator, Pointer, Multiplex and Path. This extra layer provides SDH with a different mapping capability from SONET.

Some terminology differences exist between SONET and SDH for equivalent functions or elements. Table 19-3 below summarizes that comparison. For example, ATM payloads are mapped into SONET STS-1 or STS-3c signals. As discussed above, SDH payloads are mapped first into an Administrative Unit such as AU-3 and AU-4, before being mapped into an STM-1. The SDH, the TU is equivalent to the SONET VT. In SDH, the payload is contained in the Virtual Container whereas, SONET places it in the Synchronous Payload Envelope.

However, SDH and SONET differ in how they map lower-speed digital signals into the synchronous frame and in maintenance overheads. SONET provides one mapping path for each payload type, while SDH provides

SONET	SDH
STS-1	AU-3
STS-3c	AU-4
VT	TU
VT Group	TUG-2
VT1.5	TU-11
VT2	TU-12
VT6	TU-2
VT SPE	VC
VT1.5-SPE	VC-11
VT2-SPE	VC-12
VT6-SPE	VC-2
STS-1 SPE	VC-3
STS-3c SPE	VC-4
STS-12c SPE	VC-4-4c
STS-48c SPE	VC-4-16c

Terminology Comparison SONET - SDH
Table 19-3

at least two alternate paths for each payload. Thus, interoperability could present the need for compatibility planning between SDH and SONET systems.

For example, an ATM payload at 149.76 Mbps can be transported on a SONET STS-3c SPE. This envelope maps to an SDH envelope of VC-4 for an AU-4 based STM-1. In this example, SONET and SDH Network Elements use compatible mappings which would make it possible to send the SPE transparently between SONET and SDH systems. Although the mappings at this level are compatible, conversion capability is needed to adapt overhead differences along with the passing of the SPE. Some SDH and/or SONET NEs offer interface cards for both systems (for example STM-1 and OC-3). If the two mappings are not compatible between the internal multiplex structure of their two signals (STM-1 and OC-3), the differences would have to be worked out. This hypothetical SONET/SDH interface is shown in Figure 19-1.

Both SONET and SDH are designed to accommodate timed networks and synchronous multiplexing for efficient payload entry, exit and placement with the multiplexed stream.

Overall, for compatible SONET and SDH mappings, there are no discrepancies between transport formats for carrying ATM payloads. However, there are some differences in their overheads, interpretation and mapping. Also, there is no discrepancy between the transmission rates, although the basic SDH format begins at three times the SONET basic frame rate. Much effort has been made to bring SONET and SDH into compatibility. Depending on the payloads and levels no conversion may be needed. However, differences exist in some areas. Examples are shown in Table 19-4.

Table 19-5 is a summary of the SONET and SDH signal definitions and relevant transmission services.

Area	Discrepancy Examples Between SONET-SDH
Transport Overhead Format	Nomenclature: • SONET Section overhead is SDH Regenerator Section overhead • SONET pointers in Line overhead are AU-n pointers in SDH • SONET Line overhead except pointer is Multiplex Section overhead in SDH
Section/Regenerator Section Trace; J0	When a 16 byte trace message format is used in SDH, the SONET system can not interpret this message.
Synchronization Messaging/Status	A discrepancy exists between SONET and SDH regarding this byte.
Tandem Connections/ Network Operator Byte: N1	Tandem Connection definition, used in SDH, is considered optional and not normally used in North America, therefore this byte is not applicable to SONET.

Some Differences SONET - SDH

Table 19-4

SONET	SDH	SPE Capacity	Payload (Service)
VT1.5 - SPE	VC-11	1.664 Mbps	1.544 Mbps (DS1)
VT2 - SPE	VC-12	2.240 Mbps	2.048 Mbps (E1)
VT3 - SPE	N/A	3.392 Mbps	3.152 Mbps DS1C
VT6 - SPE	VC-2	6.848 Mbps	6.312 Mbps (DS2)
OC-1	N/A	51.849 Mbps	44.736 Mbps (DS3) Note: DS3 can be supported in SDH on a VC-3/STM-1
OC-3	STM-1	155.520 Mbps	149.76 Mbps (ATM) or 139.264 Mbps (E4) or 134.208 Mbps (3 DS3)
OC-12	STM-4	622.080 Mbps	599.040 Mbps (ATM)
OC-24	N/A	Obsolete	N/A
OC-48	STM-16	2.488 Gbps	2396.16 Mbps (ATM)
OC-192	STM-64	9953.28 Mbps	9584.64 Mbps (ATM)

VT = Virtual Tributary

STM = Synchronous Transport Module

OC = Optical Carrier (a SONET Optical Signal)

EC= Electrical Carrier (a SONET Electrical Signal)

Concatenated = Continuous pipe / linked together (e.g. STS-3c)

Note: VT2 is a SONET Virtual Tributary operating at a Payload Capability of 2.048 Mbps. However this is used for E1 transport, which is common internationally but not common in North America. Therefore it is not included in the North American summary of speeds and capacities.

Note: DS0 and DS3 payloads are transported on SONET common. However, the vehicles to do this in SONET are STS-1 and VT frames, respectively, which are of higher bit rate. These SONET frames can multiplex several signals together as appropriate and add overhead.

SONET and SDH Payload Definitions Summary
Table 19-5

PART 3

Applications and Standards

20 ATM APPLICATIONS

ATM technology is used to transmit information across telecommunication networks and computer networks. Figure 20-1 ATM Applications, illustrates the types of applications that can be carried by ATM network and to whom those applications are geared. Major ATM applications include client-server applications, large image-based file transfer, real-time multimedia applications and disaster recovery.

ATM's ability to support circuit and packet mode information, to provide end-to-end bandwidth and to Quality of Service (QoS) guarantees uniquely enables support of these disparate applications. QoS guarantees allow for multiple service types to gracefully coexist with each receiving their optimized use of network resources. Further, ATM's multi-service capability allows new ATM-enabled applications and wide-area services to be integrated with existing non-ATM based services. This is accomplished in the same network while under the control of ATM-based network management. With the burgeoning bandwidth growth among disparate networks, ATM integrates voice, data, and video onto a single platform. This integration greatly benefits service users.

20.1 PERMANENT VIRTUAL CONNECTION (PVC) SERVICES

The first Broadband services introduced were Permanent Virtual Connection (PVC) based applications such as Classical IP over ATM, PVC Cell Relay Service (CRS), Frame Relay Service (FRS), and the PVC version of Switched Multi-megabit Data Service (SMDS). PVCs as

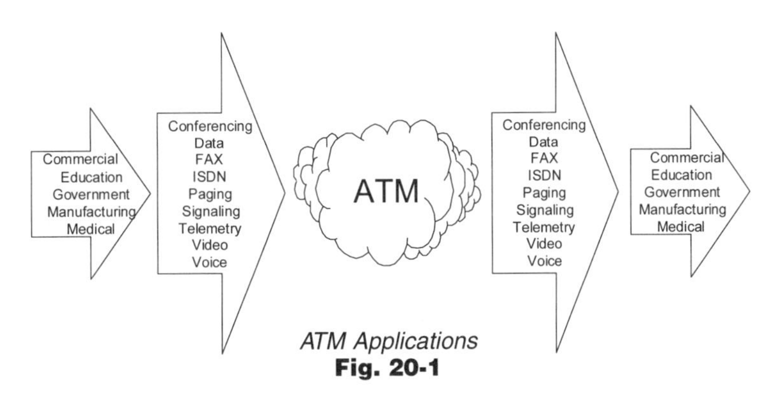

ATM Applications
Fig. 20-1

Private Line DS1/DS3 services can also be supported using Circuit Emulation Service. These PVC services do not require functions related to terminating call control signaling, performing real-time call management, providing traffic management and policing functions. The following briefly summarizes PVC services commonly available:

• *Classical IP Over ATM* [14 & 15] — Classical IP Over ATM RFC 1577 standard and the Multiprotocol Encapsulation over ATM Adaptation Layer 5 RFC 1483 were developed by the Internet Engineering Task Force (IETF). The objective was to enable existing IP applications to run transparently over ATM with no modifications, and to allow ATM hosts to transparently communicate to existing IP hosts (e.g. Ethernet). Classical IP Over ATM is supported at any bandwidth rate available up to the maximum peak bandwidth of the UNI. However, early implementations have revealed that TCP windowing flow control limits (typically hitting the maximum number of packets that can be sent before acknowledging receipt of the packets) are reached well before the maximum UNI wire speeds can be achieved. This is being resolved as vendors optimize their TCP implementations.

• *PVC-based ATM Cell Relay Service (CRS)* - Cell Relay is the native fast-packet service of ATM-based B-ISDN. It is a connection-oriented, cell based information transfer service providing wide-area, high-speed information transfer among distributed locations, and is designed to support a mix of applications, including data, video, image transfer, and multimedia. CRS can support connections with peak rates up to the maximum wire speeds of User Network Interfaces (UNI). Currently the highest rate UNI defined is 622 Mbps.

• *PVC-based Frame Relay Service (FRS)* [17] - Frame Relay is a 64 Kbps to 1.5 Mbps connection-oriented data service designed to support LAN interconnection and terminal-host applications.

• *Switched Multi-megabit Data Service (SMDS)* [16 & 18] - SMDS is a Public Packet Switched Data (PPSD) service that provides for the transfer of variable-length data units at high speed (initially up to DS3 without ATM and up to the maximum rate of ATM UNIs) without the need for call establishment procedures. It supports applications such as LAN interconnect, data transfer and image transfer.

• *Private Line or Circuit Emulation Service* [19] - This service will support interoperability between the ATM-based B-ISDN and the existing embedded base of equipment developed to operate using DS1 or DS3 connections between locations. Circuit Emulation Service (CES) defines how the DS1 and DS3 signals are encapsulated and carried over ATM transparently.

The provisioning of PVC services, while satisfactory for small numbers of customers, is not practical when many thousands or millions of users must be supported by the network. PVCs require the service provider/operator to manually negotiate the characteristics of the service, expectations for quality of service, and guarantees,as well as penalties if those service guarantees are not met. This process is referred to as Service Level Agreements (SLAs). The iterative process associated with SLAs is illustrated in Figure 20-2. Given the rate at which user/corporate network customer traffic is changing, the SLA with

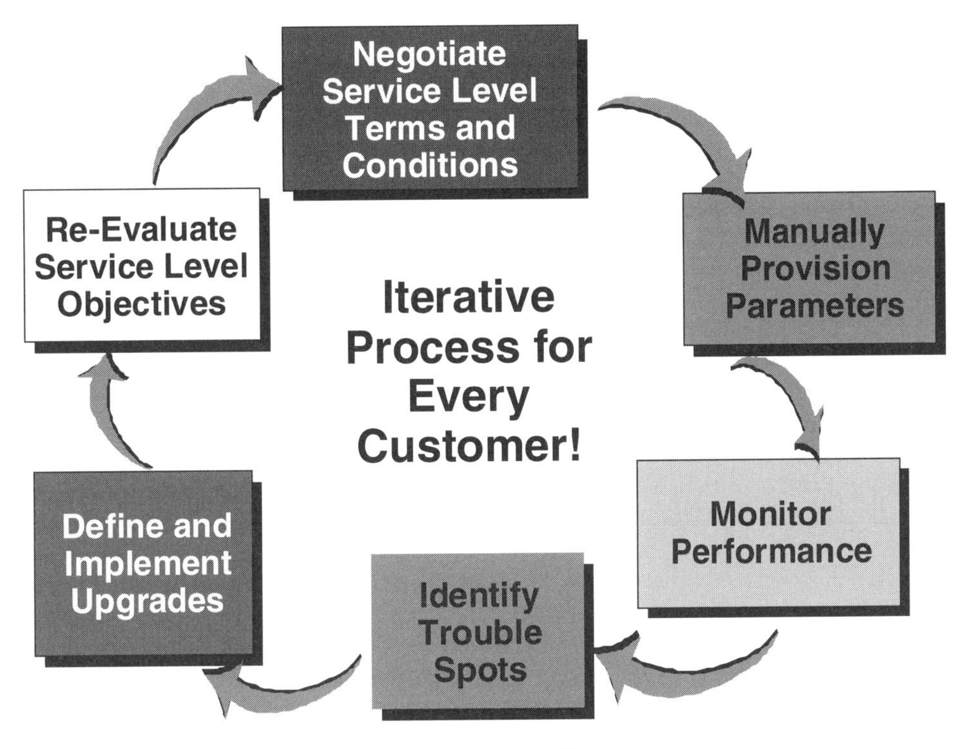

Service Level Agreement (SLA) Process
Fig. 20-2

that customer must be re-negotiated periodically. Consequently, SLAs require a good deal of human resources and they are prone to errors. Switched Virtual Connection (SVC) based services automate this process, saving time, resources and errors.

20.2 SWITCHED VIRTUAL CONNECTION (SVC) SERVICES

While PVC services were introduced into the network first, the ultimate objective is to offer SVC services. SVC services offer the potential for lower cost since users are charged only when connections exist. The introduction of the signaling [7, 8, 9 & 10] enables applications to gain access to more of the capabilities that ATM offers and allows added value network services to be provided.

SVC services provide more opportunities and apply to more environments. SVC services requirements and specifications are divided into the following categories for discussion in this section:

- Integrating Voice and Video: Multimedia
- Video Dial Tone and Conferencing Service
- Interworking existing LANs and IP/Internet
- Frame Relay (FR)
- Campus/Corporate Enterprise Networks

20.3 INTEGRATING VOICE AND VIDEO

Standards and industry implementation agreements to support voice and video were only recently completed. After the original 13 ITU-T ATM concept Recommendations were approved, priority was placed on adapting data applications to ATM because of greater need and opportunity. The first ITU standards to take voice coding and adapt it to ATM were completed in 1996. The ATM Forum built on these international standards and, in turn, developed the Voice and Telephony Over ATM (VTOA) to the Desktop [20], and VTOA ATM Trunking Using AAL1 for Narrowband Services [21].

VTOA Desktop defines the use of AAL 5 operated in the CBR mode to carry voice and provide timing information. The use of AAL 1 is optional[18] However, within international standards, the situation is reversed. AAL 1 is preferred and use of AAL 5 is optional. Interworking between AAL 1 and AAL 5 is also addressed in the VTOA Desktop specification. [Additional aspects such as signaling and control of the VTOA Desktop application are specified in conjunction with UNI Signaling version 4.0, ITU-T Q.2931 Access Signaling protocol which includes interworking between Broadband and Narrowband ISDN networks.]

The majority of added value/supplementary services have been defined and developed for narrowband networks, however, they have not been extended or adapted yet for ATM Users. These adaptations are just beginning in the industry. New work is also underway on VBR/packet voice over ATM.

VTOA Trunking specifications define, in more detail, interworking between ATM Broadband and ISDN Narrowband networks. The specification defines several reference configurations for Interwork Function (IWF) unit and specifies how information from user plane, signaling and control plane, and Operations Administration and Management (OA&M) functions are handled and interworked.

Multimedia capabilities are defined in the ATM Forum's Audio-visual Multimedia Services (AMS) [22] specification. This specification addresses the carriage and the signaling and control of audio, video, and data over ATM. Phase 1 specification currently addresses the service requirements for video coding using Constant Bit Rate (CBR), Packet Rate (CPR), and MPEG-2 (Motion Picture Image Coding Expert Group Level 2) [23] program transport stream support. The areas specified establish requirements for AAL 5 CBR mode of operation, the encapsulation of MPEG-2 Transport Streams into AAL 5 PDUs, signaling and connection control requirements, traffic characteristics and Quality of Service (QoS) characteristics and requirements.

At this time, the specifications rely on Constant Bit Rate (CBR) connections and use ATM Adaptation Layer Type 5 for both CBR Voice and CBR Video. The use of CBR connections simplifies the adaptation protocol functions associated with timing, however, the ultimate goal is to operate over Variable Bit Rate (VBR)/packet connections. Migrating towards VBR enables more efficient transport based on

---

[18] VTOA Desktop objective is to support voice in the Variable Bit Rate (VBR) or packet mode. However, version 1.0 of the specification defines the use of AAL 5 operating in the CBR mode. Currently there is no industry consensus on how to communicate timing information necessary to maintain speech synchronization in the VBR/packet mode. CBR mode was developed first precisely for real-time dependent applications since CBR mode provides timing from the network. However, greater network efficiency can be realized with speech encoded as VBR/packet connections. Industry efforts are underway developing VBR/packet voice standards.

packet switching. In the future, all forms of traffic, i.e. voice, video as well as existing data, will operate in the VBR/packet mode. This technology trend is the principle reason why AAL 5 in the CBR mode of communication was required rather than AAL 1. It will facilitate migration and equipment interoperability between VTOA and AMS CBR/VBR implementations with the same AAL 5 layer.

An alternative to the AMS multimedia specification being developed for packet mode communications is based on the ITU-T H.323 suite of Recommendations and the IETF. In support of this alternative, the AAL protocol functions necessary to adapt the H.323 suite of standards to ATM have been defined in *"H.323 Media Over ATM".*[19]

H.323 standard is used as a short hand notation representing a suite of approximately 16 standards for non-real time packet voice and video with audio. The suite of standards was developed to solve the problem where Internet based Voice over IP products were not interoperable. With chips becoming generally available, and interest in Voice over IP (VoIP) growing significantly, H.323 based products and terminal devices will likely become widely available in the future. This reflects the interest and shift towards all packet networks that was discussed previously in Part 1.

The H.323 suite of standards defines the functions involved in signaling and control of media connections to work over IP packets or ATM cells in the VBR mode, as well as frame formats (e.g. Ethernet), Figure 20-3. Residential H.323

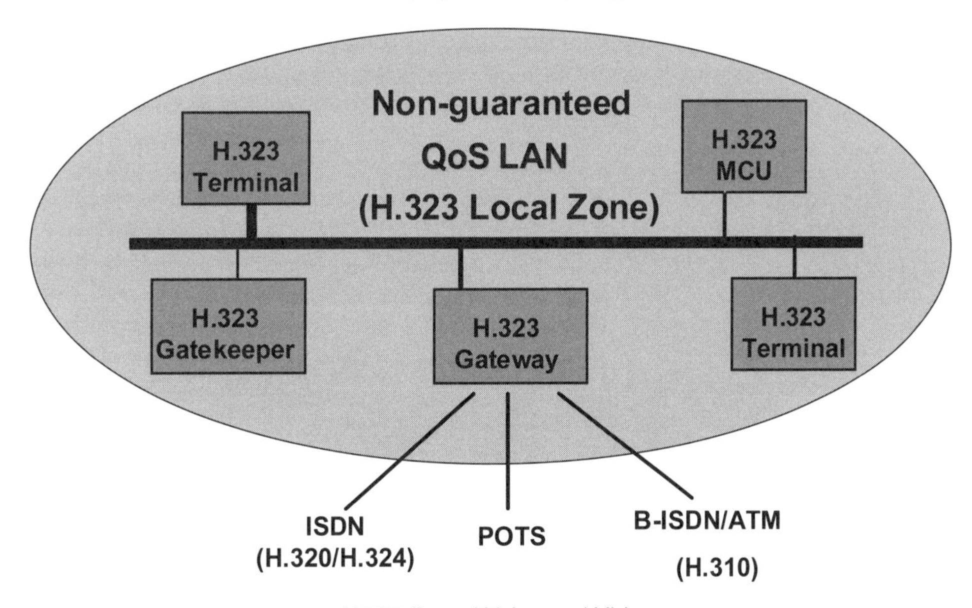

H.323 Based Voice and Video
Fig. 20-3

[19]The ATM Forum specification "H.323 Media Over ATM" is being incorporated and published in ITU-T Version 3 of the H.323 suite of Recommendations.

Gateways will allow voice calls and video connections to be supported over cable modems or xDSL modems in an integrated manner. The new and evolving text-based protocol for signaling and control is based on IP (not telephony based out-of-band signaling) called the Media Gateway Control Protocol (MGCP). It is developed to manage calls or connections from an external intelligence point, known as a Call Agent, Figure 20-4. The Call Agent is a centralized control entity located in the network and is responsible for creating the connection. Users served off a H.323 Gateway operate under the control of the Call Agent. When a packet voice call is placed, the Gateway uses MGCP to talk to the Call Agent with information such as connection parameters and the endpoint address/number of the destination party.

The Call Agent is also responsible for performing signaling interworking, if necessary, between the existing circuit-based telephony world and the packet world of IP or ATM. In the near future, computer/PC users who are IP enabled and install an H.323 Network Interface Card (NIC) and Gateway software will be able to support voice packet calls over a single line. The Gatekeeper provides not only local zone management but address resolution and admission/resource control.

If the destination is an existing telephony user, the Call Agent interworks MGCP signaling information with the Public Switched Telephone Network (PSTN) to enable an end-to-end packet/circuit call. This is essential because many of the destinations in existence today are not IP terminals and will probably continue to be "black" telephones for decades to come. The Gateway provides the protocol conversion functions needed for packet/circuit interworking at the data plane or speech coding level. Previously, Voice over IP

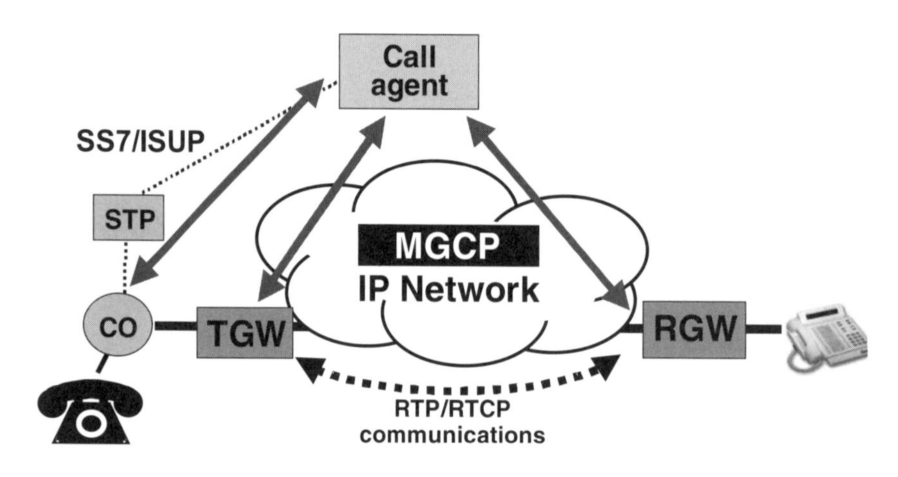

CO: Central Office, existing telephony
RGW: Residential Gateway
RTP: Real-time Transfer Protocol
RTCP: Real-time Transfer Control Protocol
SS7/ISUP: Signaling System No. 7/ISDN User Part
STP: Signal Transfer Point
TGW: Trunk Gateway

MGCP Basic Connection Setup
Fig. 20-4

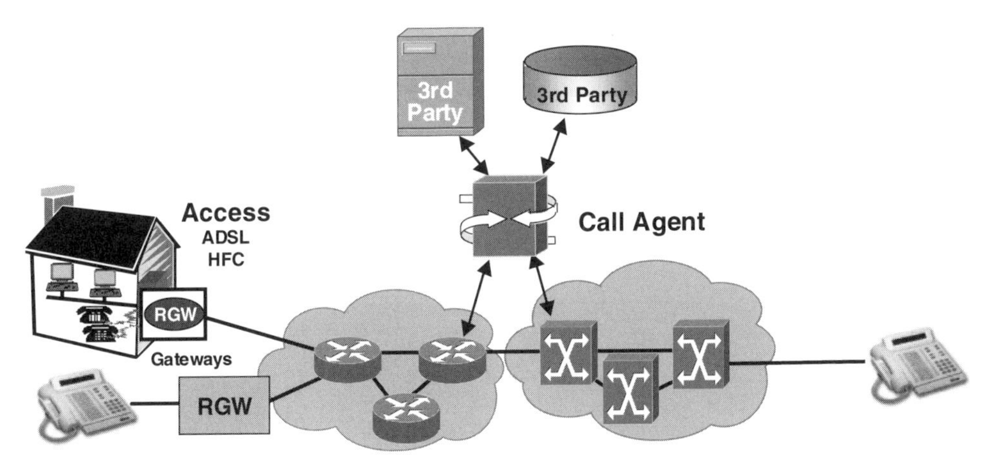

3rd Party Services Enabled with MGCP
Fig. 20-5

(VoIP) calls could only be made to another party connected on the Internet and could not be made to the vast majority of telephone users served by the PSTN. The MGCP protocol is crucial for equipment vendors developing products for the packet based Next Generation Network (NGN).

MGCP is technically significant because it introduces new control plane capabilities into the data centric Internet to support high value voice services. Further, it transitions the voice network fabric from circuit to packet (ATM and IP), and separates intelligence from the embedded switching and transport systems, Figures 20-4 and 20-5. The Call Agent with this new signaling and control solution will enable innovative service software to be provided by third parties more rapidly than what has been achieved in the past with PSTN based intelligent networking.

Residential Broadband: At this time, work is not complete regarding ATM based access distribution arrangements for residential broadband applications. The ATM Forum is developing technology independent access requirements upon which other organizations can base their technology specific approaches. For example, there are efforts in the DSL Forum (Fromerly known as the Asymmetric Digital Subscriber Line (ADSL) Forum) on ADSL/ATM Mode implementation agreements for twisted wire pair lines. Other related standards activities includes ITU developing a standard referred to as G.lite, similar to ADSL but does not require splitters to be installed on the 2-wire copper loop, and IEEE 802.14 for cable modems served by a Hybrid Fiber Coax distribution network. It is interesting to note that G.lite is based on use of ATM, Today 80 percent of the ADSL implementations also are based on ATM. In early 1999, the DSL Forum decided to base all of its work (ADSL, G.lite, SDSL, HDSL, etc.) assuming ATM solution. It is expected that the xDSL solutions will become 100 percent based on ATM in the near future.

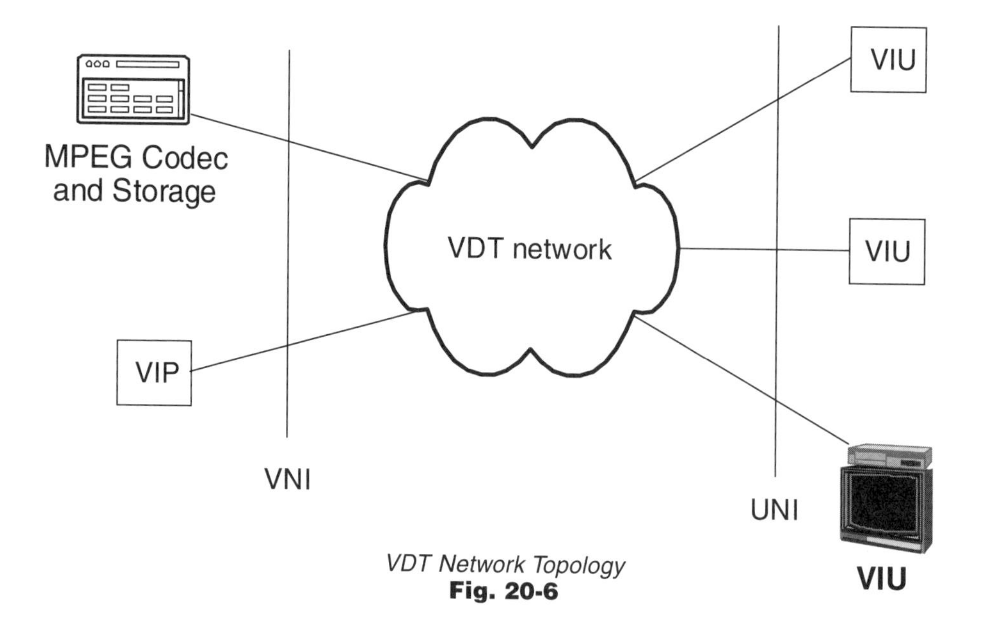

VDT Network Topology
Fig. 20-6

20.4 VIDEO SERVICES

20.4.1 Video Dial Tone (VDT)

Video information usually is compressed in digital formats for efficient storage and transport due to the nature of its large file size. For example, the most common application of Motion Picture Image Coding Expert Group Level 2 (MPEG-2) video needs a bandwidth from 2 Mbps to 6 Mbps. Video Dial Tone (VDT) services have been used for common carrier network services to provide end-users access to a variety of video information. The major components of VDT network architecture include a VDT network, Video-to-Information Providers (VIPs) and Video Information Users (VIUs). See Figure 20-6, where VNI is the

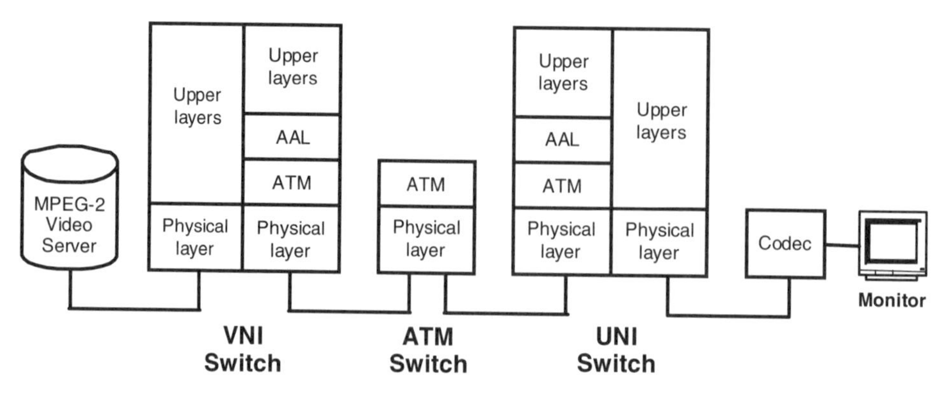

End-to-End VDT Network Architecture
Fig. 20-7

VIP-Network Interface and the UNI is the user interface. VIPs include cable TV providers, advertisement agencies and educational TV stations.

To provide real-time access to the video applications, an ATM network can be used as the VDT network or part of the VDT network. Figure 20-7 illustrates the protocol implementation using ATM as the VDT network, where the ATM switch represents the ATM network. The layers higher than the ATM layer are transparent to the ATM network, meaning that the higher layers are carried over the ATM network without any processing.

20.4.2 Video Conferencing and Distance Learning

Business video conferencing and distance learning applications can be supported over ATM using equipment that is becoming widely available. Two different types of Codecs are available for these applications, MPEG-based, which was previously discussed, and JPEG (Joint Picture Expert Group). JPEG was originally developed for narrowband network connections, but the shift is towards broadband and MPEG based connections to provide better quality. Unfortunately, there is no effort at this time to interwork among broadband and narrowband video conferencing coding solutions involving MPEG and JPEG respectively.

Conference sessions can be established in either a decentralized manner or in a centralized manner under control of the Network Controller, Figure 20-8. The distributed manner is far more convenient for the user. Conference sessions are set up on demand with the use of SVC signaling. The centralized mechanism utilizes a network controller that can be pre-programmed to establish conference connections on the behalf of the users at a particular time. This technique,

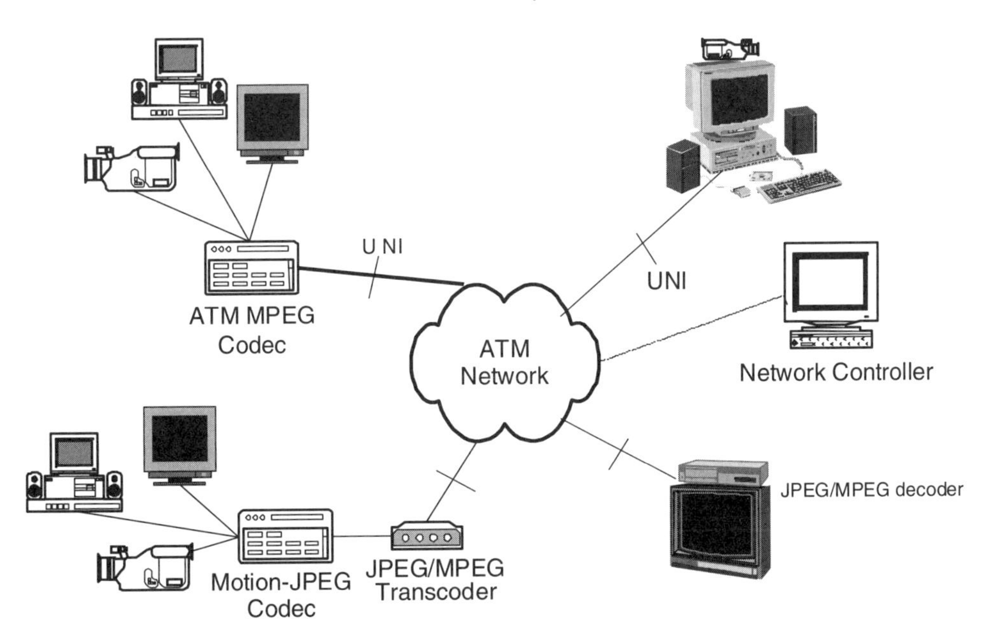

Video Conferencing and Distance Learning
Fig. 20-8

widely used in narrowband network implementations and for distance learning/education environments, requires a separate protocol for communications and programming of the network controller.

Conferencing and distance learning implementations require a multimedia computer/PC device with sound card, microphones, speakers, and appropriate software. A NIC (Network Interface Card) connects the terminal to the local network access wiring and one or more multimedia communications servers. The NIC in the desktop device would need to support one of two possible sets of standards: VTOA Desktop with MPEG/JPEG video coding and/or the AMS specification, or alternatively the H.323 suite. With computers/PCs that are already IP/Internet enabled, it is likely that the H.323 suite of packet voice and video will emerge as the preferred long-term solution. Terminal middleware would involve the Native ATM Services: Semantic Description version 1.0, defining Application Program Interfaces (APIs) access to ATM capabilities. This API driver enables native ATM applications and other service components within an end system to access ATM capabilities and services. The ATM Forum has worked collaboratively with the leading sources of operating systems to adapt their OS-dependent APIs to ATM. This includes WinSock 2 for the Microsoft environment, XTI with the Open Group consortium for the Unix environment and AIW (APPN Implementers Workshop) for the IBM environment. With these specifications, existing applications will operate transparently over ATM while enabling the development of ATM aware applications.

If multimedia applications are to be successful, it must become easier for users to connect their equipment to networks and activate service. The H.323 approach, based on the IP protocol, will enable an icon driven interface thereby making it easier for users. Service providers and/or carrier networks will also need to support the MGCP protocol, discussed earlier, and the Integrated Local Management Interface (ILMI) [28] to achieve this goal.

20.5 INTERNETWORKING LANS WITH ATM

There is a large installed base of Local Area Network (LAN) solutions. Integrating ATM into an existing local environment to leverage current investment, experience, and large numbers of applications requires that ATM technology support features in legacy LANs and end-systems in a transparent manner. However, there are several technical differences between traditional LAN technologies and ATM. These include:

- LAN technologies, such as Ethernet and Token Ring, are connectionless packet-oriented, while ATM is connection-oriented.
- The packet sizes of LANs vary up to 1500 octets for Ethernet and even higher with other LAN standards. ATM has fixed size cells or packets of 53 octets.
- LANs operate as best-effort mode. No guarantees are provided. If information delivery is important, users implement higher layer protocol processing to provide assurance of delivery. ATM has developed equivalent traffic management mode of operations offering Unspecified Bit Rate (UBR) and Available Bit Rate (ABR) service to match IP “best effort” communications and offer

a higher class of service respectively.

- Finally, LANs use broadcast and group Media Access Control (MAC) addresses. ATM equivalent functionality has been developed using point-to-point connections between servers.

There are two common approaches to ATM and LAN internetworking with the existing LAN protocols. The IETF developed Classical IP Over ATM and the ATM Forum developed LAN Emulation (LANE) [24]. Both approaches achieve the objective of existing applications to operate transparently. However, these two approaches achieve this at different layers in the protocol stack and consequently offer somewhat different capabilities. Both are briefly discussed.

20.5.1 Classical IP Over ATM

"Classical IP Over ATM", RFC 1577, is a native mode protocol that defines how to support IP directly over ATM. It operates at the network layer i.e. the IP layer in a TCP/IP system implementation thereby treating ATM as another link layer. The methods for transmitting IP over ATM are defined in the following series of RFCs: 1483, 1577 and 1626. RFC 1483 *"Multiprotocol Encapsulation over ATM Adaptation Layer 5"* describes how encapsulation is done as well as defines how packets may be multiplexed on the same connection. (In spite of the title, it does not define the support of other protocols over ATM. This was done by the ATM Forum LAN Emulation specifications which are discussed in the next section.) RFC 1483 defines the use of a LLC/SNAP packet prefix or VC multiplexing. RFC 1626 *"Default IP MTU for use over ATM AAL 5"* defines the maximum transfer unit (MTU) size when transporting IP packets over ATM as 9180 Bytes. This size was chosen to align with the MTU size for IP over SMDS.

RFC 1577[20] *"Classical IP over ATM"* defines a method for automatic resolution of IP addresses. Instead of doing a MAC to ATM address mapping as defined in LANE, IP over ATM maps IP addresses into ATM addresses. Classical IP over ATM registers its own address with an address server located in the ATM network. The ATM adapter with IP over ATM protocol uses the server to learn the ATM addresses of other Classical IP over ATM devices within the network. This differs from traditional IP where devices use Address Resolution Protocol (ARP) broadcasts to learn remote addresses. Since Classical IP over ATM does not support broadcast or multicasting, an IP over ATM device must rely on ATM address resolution services explicitly, and not the traditional LAN broadcast/response mechanisms.

The advantage of using RFC 1577 is that it has the potential to deliver better throughput than LAN Emulation can offer due to having less overhead and a larger maximum packet size. However, RFC 1577 uses a single ARP server. This network element may become a performance bottleneck and also may represent a single point of failure. This severely limits the scalability of the system because the single server can become a performance bottleneck. Experience has shown that it represents a critical failure point as stations are

[20] RF 1577 and 1626 are obsolete and have been replaced with RFC2225 "Classical IP and ARP over ATM." Once the ATM VC is put into place between the host and the ATM destination (over which the IP packets flow), RFC 1483 would be used to encapsulate the IP packets before they are segmented by AALS.

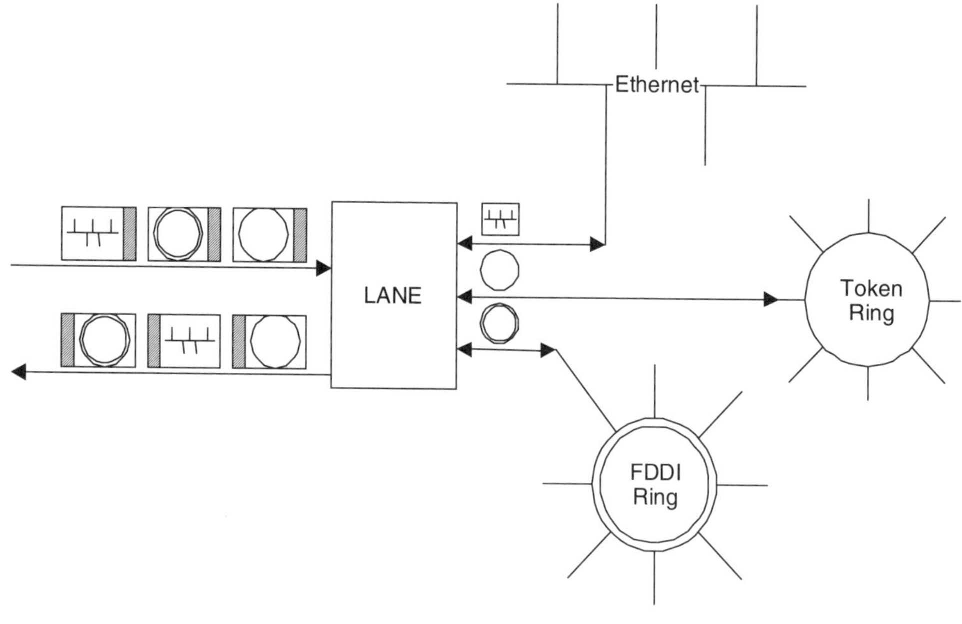

LANE Concept
Fig. 20-9

added to the network. Another disadvantage of RFC 1577 is that it is a solution for IP only. Most enterprise or corporate network environments have multiple networks[21] involving several different protocols.

LAN Emulation (LANE) is an alternative method of internetworking between an ATM network and a LAN of other network types such as Ethernet, Token Ring, FDDI. LANE encapsulates other network layer packets (e.g., IP) into ATM cells, and the mirror image process takes place at the other transmission direction (See Figure 20-9: LANE). LANE is a mechanism to carry low-speed network protocols over a high-speed ATM network. It does not, however, utilize the QoS features of ATM.

LANE was developed to enable existing LAN technologies to run over ATM and take advantage of the higher speeds ATM provides transparently to the higher protocol layers. LANE provides use of ATM as a backbone to interconnect and extend the reach of existing legacy LANs. LANE specifies how end stations communicate with each other across an ATM network, and how ATM attached devices/servers communicate with devices on an Ethernet or Token Ring segment. LANE is a layer 2 bridging protocol operating at the Media Access Control (MAC) sub-layer of the protocol stack that causes the ATM connection-oriented network to appear to the higher protocol layers and applications as a

[21] The average USA corporation in 1999 has six (6) networks, voice, fax, and IP/Internet access, being the most common. Frequently multiple IP/Internet connections are utilized because of IP's inability to provide guarantees or quality of service separation between the different types of IP traffic, e.g. engineering from finance or human resources functions. Other mission critical applications may use protocols such as IPX, SNA, NetBEUI, etc.

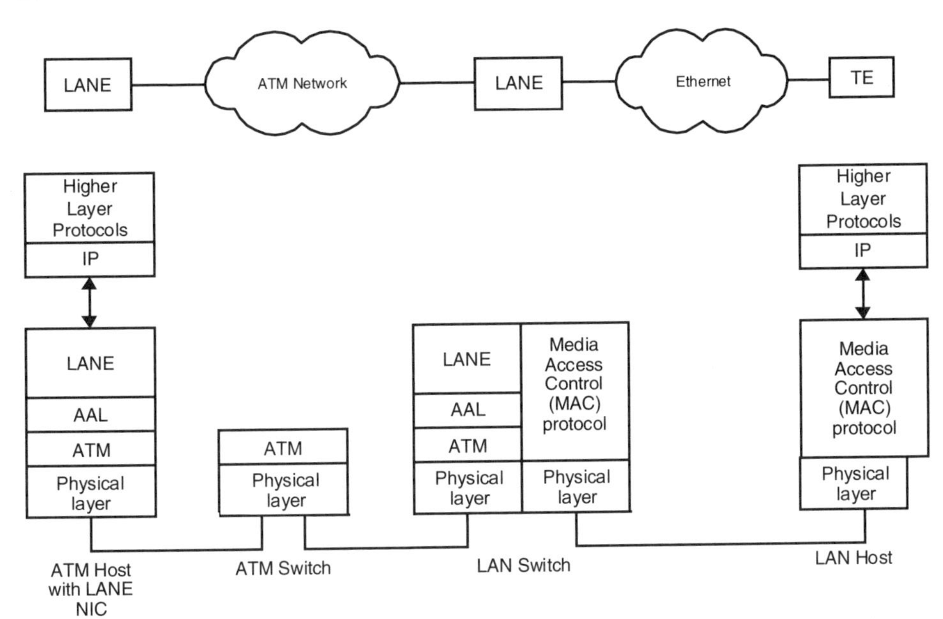

ATM-LANE-Ethernet Architecture
Fig. 20-10

connectionless Ethernet or Token Ring LAN segment. By operating at this layer of the protocol, LANE emulates a single LAN segment providing the connectionless broadcast service needed by the IP network layer protocol. It performs the necessary data conversion between LAN packets and ATM Cells. LANE does not need to emulate the Carrier Sense Multiple Access with Collision Detection (CSMA/CD) for Ethernet or token passing for Token Rings.

Different types of protocols use different standards to implement LANE. Figure 20-10 illustrates the relationship of protocols. The higher layer protocols are applications or user data. After the Ethernet protocol processing, the protocol is mapped into the ATM cell at the LAN switch with LANE capability. When the cells are transferred over the ATM network, the network only processes the ATM layer and the Physical layer information. The layers higher than the ATM layer are transparent to the ATM network; i.e., the higher layers are carried over the ATM network without any processing. The cells are recovered as Ethernet protocols at an ATM switch with LANE capability before they arrive at their destinations.

LANE provides several capabilities not possible with Classical IP over ATM. The LANE approach solves the problem of a device discovering the MAC address of the destination station. This is done by having clients registering their MAC address with the LAN Emulation Server (LES) which maps the MAC address to an ATM address. LANE also provides broadcast functions, offers Virtual LAN (VLAN) capabilities, and with the LANE version 2.0 specification [25], supports redundancy as well as replicated services. This eases the network manager's job of upgrading and/or doing maintenance on servers and routers without interrupting user activities.

In addition, LANE 2.0 has capabilities to work with and support Multiprotocols Over ATM (MPOA) [26] for higher performance campus and wide area networking beyond what traditional LAN networking can achieve. Finally, LANE can accommodate multiple protocols. It is not limited just to IP. LANE supports routable protocols such as Novell IPX, DECnet, as well as non-routable protocols (such as NetBEUI, Systems Network Architecture (SNA) and its successor APPN).

Figure 20-11 illustrates the various components making up a LANE configuration. The LANE model is based on a client/server approach where multiple LANE Clients (LECs) connect to LANE servers. LANE defines three types of servers:

- LANE Server (LES)
- Broadcast Unknown Server (BUS)
- LANE Configuration Server (LECS)

While these servers are shown as separate servers for illustration purposes, they represent software components. Since the LANE specification was developed with the requirement of easy migration, most implementations combine these software server functions. This software combination occurs typically in the ATM switch or Router rather than as separate host computers. Many of the higher performance routers use ATM hardware assist to achieve higher performance. The LES is responsible for resolving MAC addresses to ATM addresses. The BUS handles flooding multicast and unicast packets with unknown destination ATM addresses among LANE Clients (LECs) on an emulated LAN. The LECs provide the functions necessary for configuring LECs with the addresses of the LES used by the attached emulated LAN (ELAN).

The LAN Emulation Configuration Server (LECS) is the central point for control and configuration of all LANE functional entities in a collection of Emulated LANs (ELANs). The LECS configures the LAN Emulation Clients (LECs) and then assigns them to a particular LAN Emulation Server (LES). In this manner, each station on an Ethernet or Token Ring has an associated LEC that is used to handle ATM transfers. For example, it is reasonable to assume that a sending device on a traditional LAN segment wants to send data to another device on another segment that may be connected over an ATM backbone network. The sending device uses its LEC to transfer/receive data over the ATM part of the path. The LEC sends a MAC ATM address resolution query containing the destination device's MAC address to the LES. The LES responds with the ATM address of the LEC associated with the target device. The originating LEC then sets up an ATM Switched Virtual Circuit, converts the MAC frames into ATM cells, and transmits the cells over the network. At the receiving destination LEC, the ATM cells are re-assembled back into MAC frames and forwarded to the receiving device.

The LECS (using a combination of ATM OA&M capabilities, server discovery mechanisms and redundancy/failure modes) detects failures in service components that cause a loss of connectivity between ELAN members. The LEC communicates with the various LESs in order to properly manage the distributed LES topology and track the status of LES instances. With this knowledge of the entire network topology, the LECS can be used to monitor network performance.

For instance, the LECS can query the status of each LES and obtain resource utilization information for load balancing purposes.

The principle function of the LAN Emulation Server (LES) is to provide address resolution. It distributes ATM addresses throughout the ELAN and transfers registered LAN destination information between LESs to enforce uniqueness of LAN destination registration on an ELAN. This allows the LESs to allocate unique IDs to the LEC which are necessary for proper communication. The LES forwards this information via control frames to other LESs over LANE Network-to-Network Interface (LNNI) control VCCs. This interface also allows notification of LES failures to allow prompt deletion of cached registration information associated with clients of the failed LES. (LANE version 1.0 defines these mechanisms but does not establish the requirements and procedures. Consequently, multi-vendor interoperability is dependent on individual vendor co-operation and testing. LANE 2.0 establishes requirements and procedures so that multi-vendor interoperability can be achieved.)

The Broadcast and Unknown Servers (BUSs) transfer ELAN membership information between themselves to allow broadcast connectivity to all LEC ATM end point addresses. The BUS-to-BUS path is used to forward data frames for broadcast, multicast and unknown traffic.

In summary, LANE enables the use of ATM as a backbone to interconnect and extend the reach of existing "legacy" LANs to meet growing performance needs. It takes care of setting up connections and shipping LAN traffic in and

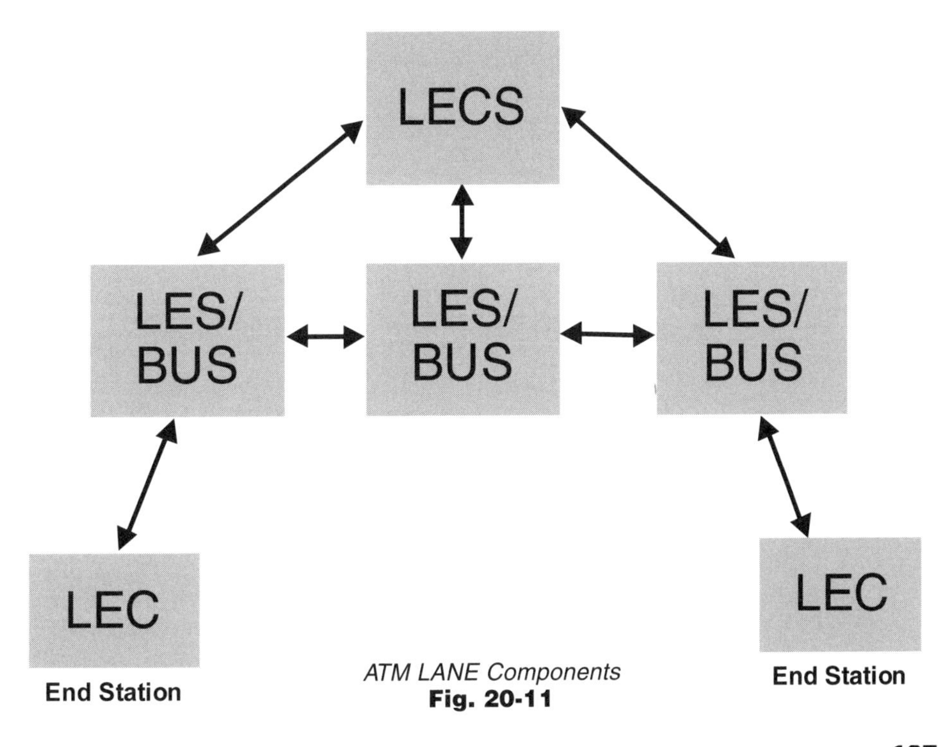

ATM LANE Components
Fig. 20-11

out of the ATM cloud/network. LANE emulates a single LAN segment by providing the connectionless broadcast service required by the IP network layer protocols. It performs the necessary data conversion between LAN packets and ATM cells and converts Media Access Control (MAC) addresses to ATM addresses.

Scaling up LANE deployment is discussed in Section 20.7 on Campus/ Corporate enterprise interworking services.

20.6 FRAME RELAY

Figure 20-11 provides a Category Type 1 interworking (as described in Part 1) example. *Frame Relay – ATM* interworking provides a method to interconnect Frame Relay (FR) networks over a high speed backbone network in such a way that it is transparent to FR customers. Other network IWF configurations can be found in ITU-T Recommendations I.555 and I.580 - "Frame Relay Bearer Services Interworking", and Frame Relay Forum (FRF) specification number FRF.5 "Frame Relay/ATM PVC Network Interworking Implementation Agreement".

The AAL layer illustrated in Figures 20-12 and 20-13 is made up of three sub-layers. In addition to the Segmentation and Re-assembly (SAR) sublayer, a Common Part Convergence Sublayer (CPCS) and the Frame Relay Service Specific Convergence Sublayer (FR-SSCS) are implicitly included.

Frame Relay-ATM service interworking represents an example of Category Types 2 and 3. In this case, FR provides a low-speed access alternative (typically at DS1/E1 rates) to ATM. FR-ATM service Interworking is used when a frame relay customer communicates with an ATM customer. The ATM customer performs no frame relay specific functions, nor does the frame relay customer perform any ATM specific functions. All of the translation and mappings from one service to the other are performed within the network and represents Category Types 2 and 3 interconnection, Figure 20-13: FR/ATM Services Interworking.

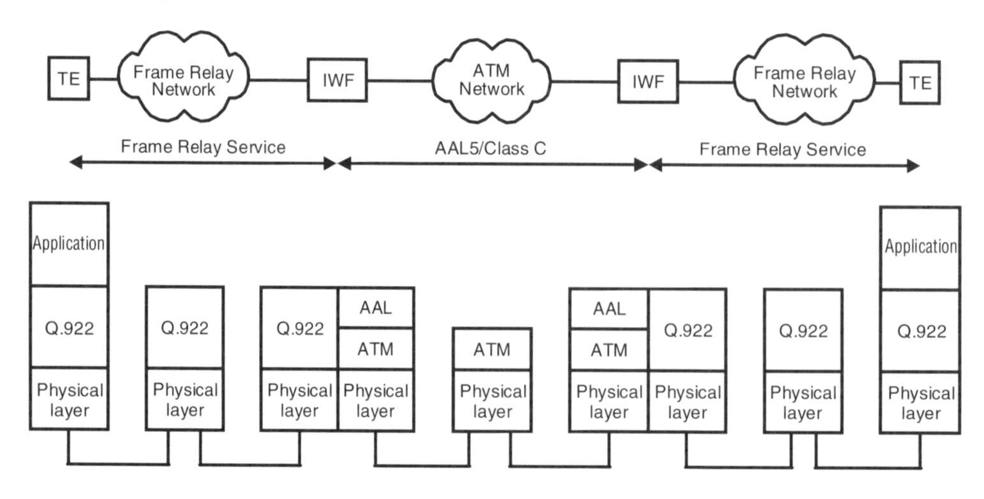

ATM/Frame Relay IWF Architecture
Fig. 20-12

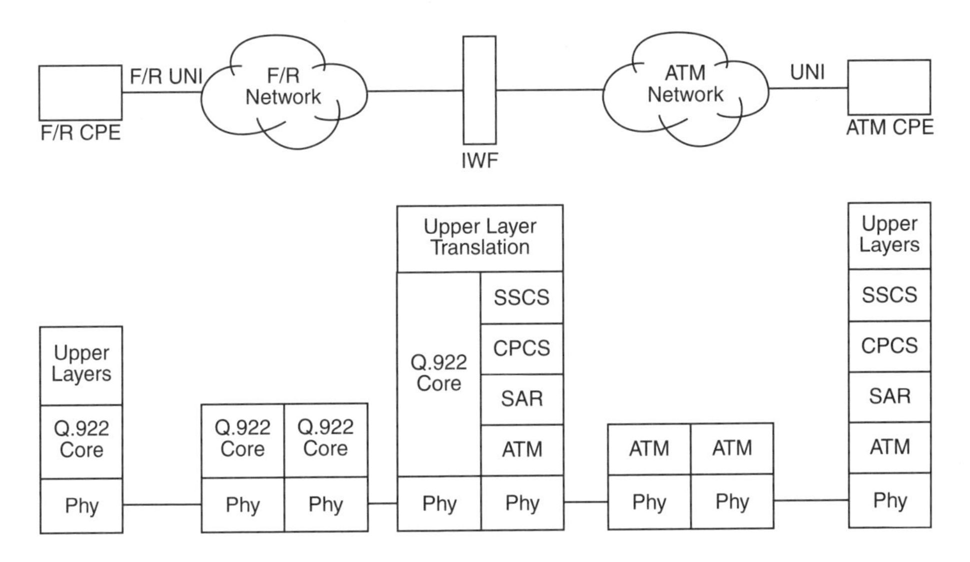

Frame Relay/ATM Services Interworking
Fig. 20-13

The need for FR/ATM service interworking comes about for two reasons. First, a corporate user may currently have an installed base of frame relay equipment, to which they are adding ATM to benefit from the higher bandwidth. A typical scenario is the case where multiple branch offices are serviced with low-speed frame relay access, while communications between the headquarters locations is serviced with an ATM high-speed access interface and backbone. Such customers will still require their frame relay users to communicate with ATM users, without requiring changes to their existing equipment and application software. Secondly, service providers/operators typically install ATM backbones in high demand/traffic areas while offering lower speed rate services in other areas. FR-ATM service interworking allows FR service to be offered on an end-to-end basis without the user needing to be concerned over what type of access they are served.

The services interworking technique was jointly developed by the Frame Relay Forum and the ATM Forum. The specification is FRF.8 "Frame Relay/ATM PVC Service Interworking Implementation Agreement."

■ 20.7 CAMPUS/CORPORATE ENTERPRISE INTERNETWORKING SERVICES

Service providers to campus and extended campus environments are continually addressing the need of escalating (and sometimes uncontrolled) bandwidth growth. These environments may require bandwidth on demand (primarily for data) via ATM for burgeoning backbones in either the campus or in the extended campus (or short-haul WAN) environments. Operators of enterprise networks or corporate wide networks prefer to focus on their core business competencies. They want to drive down the costs not associated with the core areas of their business. Campus and Enterprise operators also have needs to support a variety of different networks. All have voice networks (PBXs), many with leased or private lines. Many Enterprises have an SNA network, and most have at least one router-based LAN network and, more likely, several LAN networks, frequently using different LAN technologies based on when they were installed and if some are used as a backbone. For wide area access, most of these networks connect to the Internet and use services, such as private lines, Frame Relay, SMDS or ATM Cell Rely, to interconnect with other Campus and Enterprise networks. An ATM common platform/backbone provides a cost-effective method for satisfying these needs for escalating bandwidth among disparate networks while providing new capabilities not possible with other technologies.

Voice and Telephony Over ATM (VTOA) Desktop and *VTOA Trunking* [20 & 21 respectively] allows voice to be migrated onto the ATM platform. VTOA Trunking, combined with UNI Signaling [10], UNI Traffic Management [27], and Integrated Local Management Interface (ILMI) [28] provide the key foundation specifications to interconnect a PBX to a public carrier/service provider. Note that frequently the suite of UNI Signaling, UNI Traffic Management or UNI TM, and ILMI are generally referred to as simply, the UNI 4.0, as shorthand notation.

For the Campus and Enterprise data needs, two ATM Forum specifications are essential. They are the LAN Emulation (LANE) v2.0 LANE User Network Interface (LUNI) combined with Multiprotocol Over ATM (MPOA). Both of these specifications build on top of the UNI 4.0 related specifications (Signaling, Traffic Management, and ILMI) and are important to begin exploiting the power of ATM in the campus.

First, it is important to consider what LANE 2.0 provides beyond the basics discussed earlier in Section 20.5.2. LANE 2.0 represents the next incremental step in the migration of LANs towards a campus environment. It consists of two parts: the LUNI 2.0 (LANE UNI), a backward-compatible upgrade to a LANE 1.0 Client's interface to the LAN Emulation Service, and the LNNI 2.0 which defines the interfaces between components that go into the LAN Emulation Service. New features in LUNI 2.0 allow:

- *Multiplexing* multiple LANE Clients on Emulated LANs (ELANs) over the same VCC. This is useful for VCC constrained edge devices when there several ELANs.
- *Improved support* for QoS and mapping of IEEE 802.1p traffic classes to ATM QoS and provides support of additional ATM VCCs. Each VCC could have different Service categories. Traffic parameters can be used as Data Direct VCCs between two LANE Clients. In LANE 1.0, only one UBR Data Direct VCC was recommended.
- *Support for separating out multicast traffic* by providing the ability of a LANE Client to register for those multicast Media Access Control (MAC) addresses it wishes to receive, and thereby allow the LANE Service not to send it frames addressed to other multicast MAC addresses. Also, this feature permits the LANE service to generalize the Configuration Server so that it can apply to more than just LANE. In particular, this aspect is being coordinated with the IETF's (Internet Engineering Task Force) work on

Next Hop Resolution Protocol (NHRP). Most of these features are being optimized in such a manner in which LANE fits as one of the individual components of MPOA.

- *Association of additional control information* with the MAC address when registering the MAC address or responding to an LE-ARP. This is used by MPOA Clients/Servers to indicate that they are MPOA-capable, i.e., MPOAs auto-configuration/auto-learning is piggy-backed on top of LANE.

The LNNI is a new part of the specification, LANE 2.0. While LANE 1.0 did not preclude distributed implementations of the LANE Service components, neither did it describe how it might work. Thereby, suppliers had no choice but to go with proprietary implementations. The LNNI 2.0 not only describes how the sub-components can be distributed, but also defines the interfaces and protocols by which they exchange control and data. The LNNI's design goals are that a single ELAN can consist of 2000 LANE Clients and 20 LES/BUS Servers. Therefore, LANE 2.0, by enabling multiple distributed components, will provide a more reliable and robust local area/campus network.

Classical IP over ATM is one method of native ATM support of IP applications. It requires that the ATM equipment resolve IP addresses (as discussed earlier) and route IP packets to the next hop towards the destination using SVCs. This approach emulates traditional IP address resolution and the process must be repeated at every router hop. Further, classical IP over ATM supports only the IP protocol. Multi Protocol Over ATM (MPOA) was the first industry standardization effort to recognize per hop router performance bottleneck limitations of this classical approach, and work was initiated to develop MPOA late in 1995.

MPOA is the next piece needed for high performance campus, extended campus and enterprise network applications. MPOA is essentially a native mode (i.e., ATM based) internetworking protocol that provides a framework for effectively synthesizing bridging (layer 2) and routing (layer 3) with ATM in an environment of diverse protocols, network technologies and Virtual LANs (VLANs). MPOA expands on LANE. Unlike LANE, which merely provides data link layer 2 functionality, MPOA has a network layer interface which enables it to tap into the underlying advantages of ATM. The MPOA premise was to separate switching functions from routing functions into a multi-layer mechanism to allow application programs to gain access to Traffic Management and the Quality of Service (QoS) properties of ATM, support multiple media formats, and allow integration of intelligent VLANs.

The ability to logically group related users and their traffic on one or more campus/enterprise switches by VLANs provides users with a great deal of benefit, Figure 20-14. In addition to simplifying the add/move/change process, it also offers a greater degree of security, reliability, and traffic reduction. Perhaps most important, it facilitates direct connectivity and improved performance between attached devices (a.k.a. "shortcuts" in the form of IP/ATM switching without incurring multiple router table calculations), enables more cost effective edge devices, and provides a migration path for LANE.

MPOA utilizes a route distribution protocol developed in the IETF called NHRP (Next Hop Resolution Protocol). By separating switching from routing functions,

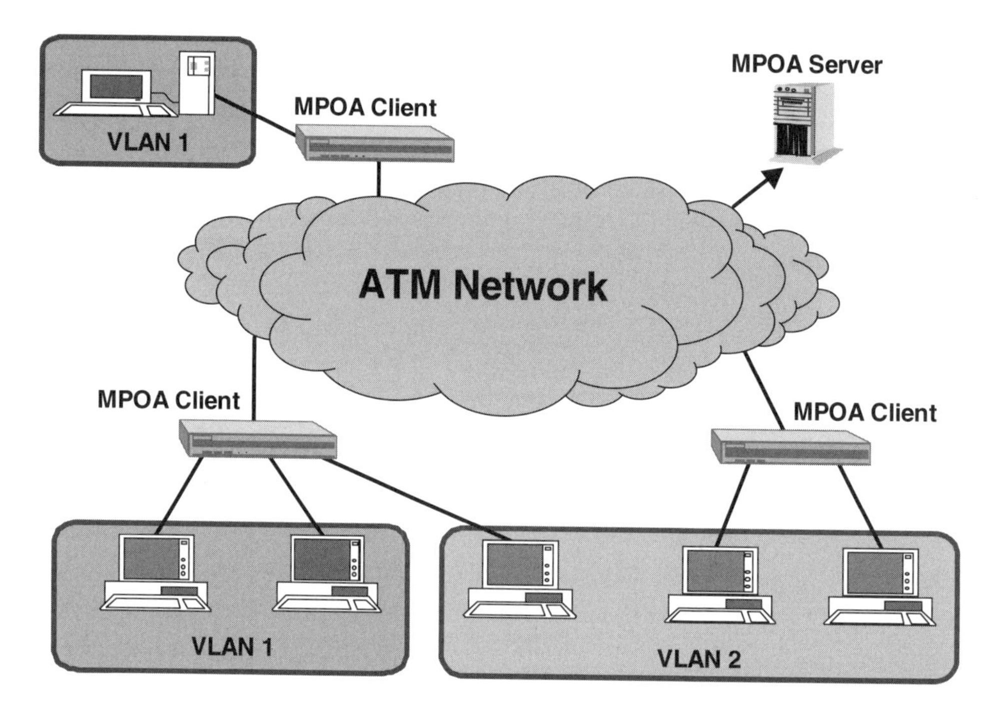

MPOA Model
Fig. 20-14

and using the route distribution protocol, MPOA provides cut-through routing/switching for more efficient transport by allowing direct, one-hop connectivity between ATM hosts and/or legacy devices that reside in different subnets. Figure 20-15 illustrates the flow of information using MPOA. If the edge device or MPOA Client knows the destination ATM address, it establishes a SVC direct or cut-through connection to that device without requiring route calculations to be performed on every packet.

If the destination address is not known, the edge device or MPOA Client will query the MPOA route Server using NHRP. If known to that server, the address is returned to the originating source which establishes the ATM SVC connection. If not known to that MPOA Server, it will query the next MPOA route server, and so on until the destination address has been determined and communicated back to the source. Once the ATM SVC connection has been established, the IP packets flow directly between devices.

The most recent addresses (or frequently accessed addresses) are cached for rapid access to minimize the number of times they have to be "looked" up. In this manner MPOA is able to avoid the hop-by-hop, packet-by-packet router calculations of a classic IP routed network.

For very short flows or when the ATM destination address are not known, IP packets flow through the routers just as in a classic IP router network.

The importance of MPOA is that it provides a multi-layer switching means to route and bridge multiple protocols over switched ATM backbones involving a maximum of three routers. It significantly reduces the need for traditional routers to handle inter-subnet traffic where the current Internet congestion problems are occurring. The Internet congestion problem has motivated the many flavors of IP Switching (a hybrid of IP routing software and ATM switching hardware) proposals that have been getting a great deal of attention.

The most prominent alternative to MPOA is the Internet Engineering Task Force (IETF) efforts on the technique called Multi Protocol Label Switching (MPLS). MPLS, started late 1996, is still being developed in 1999. MPLS' goals are similar to MPOA in that it attempts to address the router processing/calculation bottleneck for the generic Internet. MPLS currently does not provide the same degree of multi-protocol support, Traffic Management or Quality of Service provided by ATM based solutions. MPLS is being developed to use ATM, Frame Relay or narrowband leased line solutions. However, ATM is emerging as the preferred solution for many of the reasons discussed in this book. In addition, with approximately 75 percent of the Internet backbone traffic in North America now being carried over ATM[22], and with the ATM transport mechanism migrating out from the Internet backbone to the edges of the Internet, a MPLS based ATM solution will be important. In addition, MPLS added on an ATM switch and is receiving significant industry interest because it turns the switch into a router once the appropriate routing protocol is loaded.

For the present time, information Technology (IT) managers and Chief Information Officers (CIOs) who have IP/Internet router congestion have the option of going with an ATM MPOA solution, putting bigger and higher performance routers

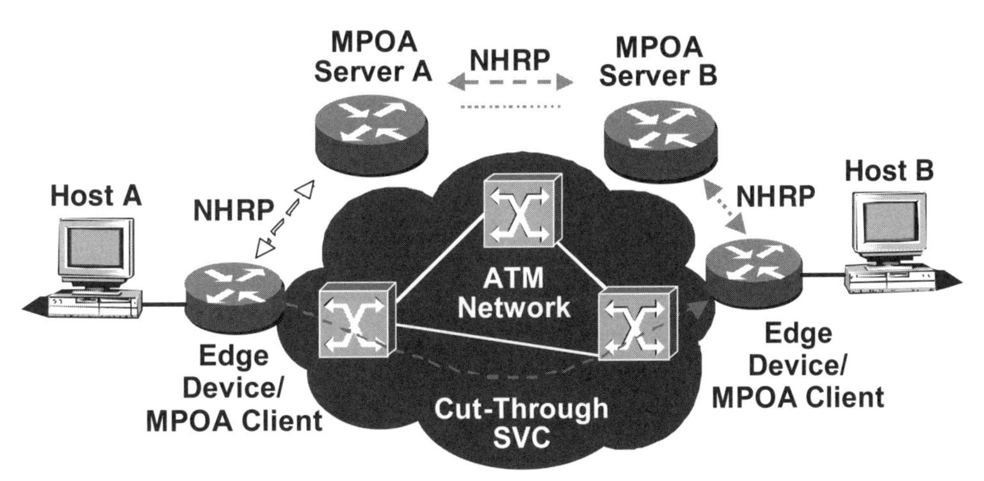

MPOA Flows – Switching vs. Routing
Fig. 20-15

[22] Personal communications between author and Mr. Fred Baker, IETF Chair.

in place of existing routers, or going with vendor proprietary versions of pre-MPLS implementations. The decision should be based on economics and careful consideration of the rapidly changing corporate communications needs. Managers face a decision since backbone components are not as easily changed as terminal equipment and servers. Once installed, these components are generally expected to have a useful life of approximately six to seven (6-7) years. Consequently, the solution must provide flexibility to meet the rapidly evolving and growing communication requirements without needing massive fork-lift upgrades, and without over engineering. ATM provides the greatest flexibility while enabling migration of other parts of the corporate network onto ATM. This migration can occur at a convenient or incremental pace with ATM's ability to provide Quality of Service guarantees, traffic management and scalability from the LAN environment to wide area network.

Summarizing some of the benefits of MPOA, it preserves the user/customer investment in legacy LANs, it solves LANE scalability issues, and provides higher performance than "full router" solution. Once a connection is established, MPOA provides performance similar to the vendor proprietary "tag switching" approach which is also being contributed into the IETF MPLS effort. MPOA provides TM and QoS capabilities to maximize network performance, and is standardized. The MPOA technique was also developed to enable application programs to gain access to the Quality of Service (QoS) properties of ATM and provide a uniform means of supporting multiple internetwork-layer protocols, such as IP (the most critical), IPX, AppleTalk, DECnet, etc. as noted previously and typically found in enterprise/corporate networks. MPOA operates on an end-to-end scaleable basis rather than hop-by-hop. Finally, MPOA provides carriers/service providers a useful value-added service tool enabling "managed" LANE or MPOA services to be offered to corporate customers.

21 STANDARDS

21.1 ATM STANDARDS

The major ATM standards organizations include the ATM Forum, the International Organization for Standardization (ISO) and International Telecommunications Union-Telecommunications Sector (ITU-T)[23]. ATM standards can be generally categorized as principles of ATM, routing protocol standards, internetworking standards, network management standards, and signaling and addressing standards. The following identifies a subset of the standards and industry specifications forming the foundation of ATM.

Principles of ATM:

- I.121: Broadband Aspects of ISDN, ITU-T
- I.211: Broadband Service Aspects, ITU-T
- I.150: B-ISDN Asynchronous Transfer Node Functional Characteristics, ITU-T
- I.361: B-ISDN ATM layer Specification, ITU-T
- I.363: B-ISDN ATM Adaptation layer (AAL) Specification, ITU-T

Routing protocol standards of significance include:

- I.362: B-ISDN AAL Functional Description, ITU-T
- Interim Interswitch Signaling Protocol (IISP), ATM Forum
- Private Network Node Interface (PNNI) Phase 1, ATM Forum
- Private Network Node Interface (PNNI) Phase 2, ATM Forum

Internetworking standards of significance include:

- I.555: Frame Relay and ATM Internetworking, ITU-T
- I.580: General Arrangements for Interworking between B-ISDN and 64 Kbps Based ISDN, ITU-T
- LANE Phase 1, ATM Forum
- LANE Phase 2, ATM Forum
- RFC 1112: Host Extensions for IP Multicasting, IETF
- RFC 1483: Multiprotocol Encapsulation over AAL 5, IETF
- RFC 1577: Classical IP and ARP over ATM, IETF

---

[23] It is formerly known as CCITT.

Network management standards of significance include:

- RFC 1157: A Simple Network Management Protocol (SNMP), IETF
- Integrated Local Management Interface (ILMI) of User-Network Interface (UNI) version 4.0, ATM Forum
- ILMI Management Information Base (MIB), ATM Forum

Signaling and addressing standards of significance include:

- I.371: Traffic Control and Congestion Control in B-ISDN, ITU-T
- Q.931: ISDN UNI layer 3 Specification for Basic Call Control, ITU-T
- Q.2931: UNI layer 3 Specification for Basic Call/Connection Control, ITU-T
- User-Network Interface (UNI) signaling Specification Version 3.1[24], ATM Forum
- User-Network Interface (UNI) signaling Specification Version 4.0, ATM Forum
- Traffic Management Specification Version 4.0, ATM Forum

21.2 SONET STANDARDS

SONET and SDH standards address standardized rates of transmission and signal formats, such as data rates in Mbps and signal framing structures including allocation of overhead. Standardization of framing and overhead significantly reduces costs and improves operational efficiency.

SONET and SDH standards also address single mode optical interface characteristics. These characteristics are the physical attributes of optical interfaces, such as the power level and pulse shape, and they include performance requirements for the optical interface.

These standards also include multiplexing schemes that result in reduced manufacturing and operating costs. The functional characteristics of network elements also are specified in the SONET and SDH standards. This is important to maintain effective, consistent and cost-effective network nodes.

The standards also address the operation of the synchronous network, including methods and guidelines for providing Stratum Level clocking to all Network Elements. This maintains a consistently high degree of synchronization in the network, greatly reducing errors from synchronization slippage or failure.

The major SONET standard organizations in the United States include American National Standards Institute (ANSI), Exchange Carriers Standards Association (ECSA), International Standards Organization (ISO). For international standards in telecommunications, the organization is the International Telecommunications Union-Telecommunications Sector (ITU-T). SONET standards were defined in three phases.

- Phase I includes transmission rates and characteristics, signal formats, optical interfaces, and the basic multiplexing structure.
- Phase II includes Operations, Administration and Maintenance (OA&M) capabilities, data communications channel protocols and ADM capabilities.

[24] Earlier versions of the UNI specification are not recommended. Backward/forward compatibility and interoperability between specification versions is only assured starting with UNI 3.1.

- Phase III includes OA&M and Provisioning (OAM&P), interconnection addressing schemes, and ring and nested protection switching standards.

These are many of the key ANSI standards relating to SONET. Additional information can be obtained by accessing the ANSI Web home page at www.ansi.org.

- T1.105 "SONET Basic Description"
- T1.105.1 "SONET - Automatic Protection Switching"
- T1.105.2 "SONET - Payload Mappings"
- T1.105.3 "SONET - Jitter at Network Interfaces"
- T1.105.4 "SONET - Data Communication Channel Protocols and Architectures"
- T1.105.5 "SONET - Tandem Connection Maintenance"
- T1.105.6 "SONET - Physical Layer Specifications"
- T1.105.7 "SONET - Sub-STS-1 Interface Rates and Formats Specification"
- T1.105.9 "SONET - Network Element Timing and Synchronization"
- T1.117 "Digital Hierarchy - Optical Interface Specifications (SONET) - Single Mode - Short Reach"
- T1.119 "SONET - Operations, Administration, Maintenance and Provisioning (OAM&P) Communications"
- T1.119.01 "SONET - Operations, Administration, Maintenance and Provisioning (OAM&P) Communications - Protection Switching Fragment"
- T1.119.02 "SONET - Operations, Administration, Maintenance and Provisioning (OAM&P) Communications - Performance Management Fragment"
- T1.245 "Directory Service for Telecommunications Management Network (TMN) and SONET"
- T1.514 "Network Performance Parameters and Objectives for Dedicated Digital Services - SONET Bit Rates"

These are some of the key standards that are used for SDH and developed by the ITU-T. Additional information can be accessed via the ITU home page at www.itu.ch.

- G.702: Digital Hierarchy Bit Rates, 11/88
- G.704: Synchronous Frame Structures Uses at 1544, 6312, 8488 and 44.736 bits/s Hierarchical Levels, 7/95
- G.707: Network Node Interface for the Synchronous Digital Hierarchy
- G780: Vocabulary of Terms for Synchronous Digital Hierarchy Networks and Equipment, 11/94
- G.783: Characteristic of Synchronous Digital Hierarchy (SDH) Multiplexing Equipment Functional Blocks, 4/97
- G.803: Architecture of Transport Networks Based on the Synchronous Digital Hierarchy, 6/97

22 REFERENCES

[1] CCITT: COM XVIII-228-E, Geneva, Switzerland, March 1984.

[2] ITU-T Recommendation I.121, "Broadband Aspects of ISDN," Rev 1., 1991.

[3] ITU-T Recommendation I.211, "Broadband Service Aspects," Rev 1., 1991.

[4] ITU-T Recommendation I.321, "B-ISDN Protocol Reference Model and Its Applications" 1991.

[5] ISO 7498-1984 (E) "Information Processing Systems-Open System Interconnection-Basic Reference Model," ATM erican National Standards Association, Inc., N.Y.

[6] C. Sunshine, "Formal Method for Communication Protocol Specification and Verification, A Rand Note," November 1979, ARPA Order No. 3460/3681.

[7] D. Minoli and G. H. Dobrowski, "Principles of Signaling for Cell Relay and FrATM e Relay," Artech House, Norwood, MA, 1995.

[8] GR-1111-CORE *Broadband Access Signaling Generic Requirements*, Issue 2, Bellcore, October 1996[25]

[9] GR-1417-CORE *Broadband Switching System SS7 Generic Requirements*, Issue 3, Bellcore, October 1996.

[10] af-sig-0061.000, *UNI Signaling* version 4.0, with af-sig-0076.000 *Signaling ABR Addendum,* (ATM Forum, June 1996).

[11] ITU-T Recommendation I.150, "B-ISDN Asynchronous Transfer Node Functional Characteristics," 1991.

[12] ITU-T Recommendation I.361, "B-ISDN ATM layer Specification," 1991.

[13] ITU-T Recommendation I.363, "B-ISDN ATM Adaptation layer (AAL) Specification," 1993.

[14] J. Heinanen, "IETF RFC 1483, Multiprotocol Encapsulation over ATM Adaptation layer 5," July 1993.

[15] M. Laubach, "IETF RFC 1577-Classical IP and ARP over ATM," January 1994.

---

[25]Note: In 1999, Bellcore changed its name to Telcordia Technologies, Inc.

[16] ITU-T Recommendation I.364, “Support of Connectionless Data Service on a B-ISDN,” 1992.

[17] ITU-T Recommendation I. 555, “Frame Relay Bearer Service Interworking,” 1993.

[18] GR-1060-CORE, *Switched Multi-Megabit Data Service (SMDS) Generic Requirement for Exchange Access and Intercompany Serving Arrangements,* Issue 1, (Bellcore, March 1994).

[19] af-vtoa-0078.000, *Circuit Emulation Service* version 2.0, (ATM Forum, January 1997).

[20] af-vtoa-0083.00, *Voice and Telephony Over ATM to the Desktop Specification,* version 1.0 (ATM Forum, March 1997).

[21] af-vtoa-0089.000, *Voice and Telephony Over ATM Trunking using AAL 1 for Narrowband Services,* version 1.0 (ATM Forum, April 1997).

[22] af-saa-0049.001, *Audio-visual Multimedia Services: Video on Demand Specification,* version 1.1 (ATM Forum, December 1996).

[23] ISO/IEC IS 13818-6/ITU-T Recommendation H.262, *Information Technology-Generic Coding of Moving Pictures and Associated Audio-Part 2: Video,* (ISO/IEC and ITU-T, 1996).

[24] af-lane-0021.000, *LAN Emulation Over ATM,* version 1.0 (ATM Forum, January 1995), and af-lane-0038.000, *LANE Emulation Client Management Specification,* version 1.0 (May 1995).

[25] af-lane-0084-000, *LAN Emulation Over ATM Version 2.0-LUNI Specification,* (ATM Forum, April 1997).

[26] af-mpoa-0087-000, *Multi-Protocol Over ATM,* version 1.0 (ATM Forum, April 1997).

[27] af-tm-0056.000, *UNI Traffic Management,* version 4.0 (ATM Forum, April 1996), with *Addendum to Traffic Management V4.0 for ABR ParATM eter Negotiation,* (ATM Forum, January 1997).

[28] af-ilmi-0065.000, *Integrated Local Management Interface,* version 4.0 (ATM Forum, September 1996).

[29] af-pnni-0055.000, *Private-Network Node Interface (P-NNI),* version 1.0 (ATM Forum, March 1996).

[30] af-bici-0068.000, *B-ICI 2.0 Addendum,* version 2.1 (ATM Forum, November 1996).

[31] GR-1110-CORE, *Broadband Switching System (BSS) Generic Requirements,* Issue 1, 1994, with Revisions through 5, October 1997 (Bellcore).

[32] GR-253-CORE, *Synchronous Optical Network (SONET) Transport Systems: Common Generic Criteria*, Bellcore, Issue 2, January 1999.

[33] GR-496-CORE, 1998, SONET Add-Drop Multiplexer (SONET ADM) Generic Criteria, Bellcore, December 1998.

[34] GR-1400-CORE, *SONET Dual-Fed Unidirectional Path Switched Ring (UPSR) Equipment Generic Criteria*, Bellcore, Issue 2, January 1999

[35] GR-1230-CORE, *SONET Bi-Directional Line Switched Ring Equipment Generic Requirements*, Bellcore, Issue 4, December 1998.

[36] ANSI T1.105.7 Telecommunications – *SONET – Sub-STS-1 Interface Rates and Formats Specification*, December 20, 1996.

[37] ANSI T1.105.2 Telecommunications – SONET – *Payload Mappings*, 1995.

[38] T1.119 Telecommunications – SONET – Operations, Administration, Maintenance and Provisioning (OATM &P) Communications, August 1, 1996.

[39] ANSI T1.105.9 Telecommunications – *SONET – Network Element Timing and Synchronization*, August 14, 1996.

[40] GR-2955 Generic Requirements for Hybrid SONET/ATM Element Management Systems, Bellcore, Issue 2, November 1998.

Other general references include:

GR-436-CORE, "Digital Network Synchronization Plan", Bellcore, Issue 1, June 1996.

FR-SONET-17, 1998 Broadband Transport Network Generic Requirements: SONET and ATM Transport Technologies, Bellcore.

GR-1374-CORE, "SONET Inter-Carrier Interface Physical Layer Generic Criteria for Carriers", Bellcore, Issue 1, December 1994.

GR-1377-CORE, "SONET OC-192 Transport System Generic Criteria", Bellcore, Issue 5, December 1998.

GLOSSARY

AAL ----------- ATM Adaptation Layer
ABR ---------- Available Bit Rate
ADM ---------- Add-Drop Multiplexer
AFI ----------- Authority and Format Identifier
AIS ----------- Alarm Indication Signal
ANS ---------- ATM Name System
ANSI --------- American National Standards Institute
ANSI/T1 ----- American National Standards Institute - Telecommunications
API ----------- Application Program Interface
APS ---------- Automatic Protection Switching
ARP ---------- Address Resolution Protocol
ATM ---------- Asynchronous Transfer Mode
BCD ---------- Binary Coded Decimal
B-DCS ------- Broadband DCS
BIP ----------- Bit Interleaved Parity
B-ISDN ------ Broadband Integrated Services Digital Network
BITS --------- Building Integrated Timing Supply
BLSR -------- Bi-directional Line Switched Ring
BOM --------- Beginning of a Message
BRI ----------- Basic Rate Interface
BSS ---------- Broadband Switching System
CBR ---------- Constant Bit Rate
CC ------------ Composite Clock
CCITT ------- Consultative Committee on International Telegraphy and Telephony
CDV ---------- Cell Delay Variation
CES ---------- Circuit Emulation Service
CLP ---------- Cell Loss Priority
CMISE ------- Common Management Information Service Element
Codec ------- Coder/Decoder
CO ------------ Central Office
COM --------- Continuation of a Message
COT ---------- Central Office Terminal
CPCS -------- Common Part Convergence Sublayer
CPI ----------- Common Part Indicator
CRC ---------- Cycle Redundancy Check
CRS ---------- Cellular Relay Service
CS ------------ Convergence Sublayer
CSI ----------- CS Indicator

DCC ---------- Data Communications Channel
DCS ---------- Digital Cross-Connect System
DLC ---------- Digital Loop Carrier
DMUX ------- Demultiplexer
DNS ---------- Domain Name Service
DS ------------ Digital Signal
DSS2 -------- Digital Subscriber Signaling No. 2
DQDB -------- Distributed Queue Dual Bus
EC ------------ Electrical Carrier
ECSA -------- Exchange Carriers Standards Association
EIA ----------- Electronic Industries Association
EOM --------- End of a Message
EOC ---------- Embedded Operations Channel
ES ------------ Edge Switch
FDDI --------- Fiber Distributed Data Interface
FEC ---------- Forward Error Correction
FRS ---------- Frame Relay Service
GFC ---------- Generic Flow Control
GFR ---------- Guaranteed Frame Rate
GPS ---------- Global Positioning System
HEC ---------- Header Error Control
IDLC --------- Integrated Digital Loop Carrier
IEEE --------- Institute of Electronic and Electrical Engineers
IETF ---------- Internet Engineering Task Force
IDLC --------- Integrated Digital Loop Carrier
IISP ---------- Interim Interswitch Signaling Protocol
ILMI ---------- Interim Local Management Interface
IP ------------- Internet Protocol
ISDN --------- Integrated Services Digital Network
ISO ----------- International Organization for Standardization
IT ------------- Information Type
ITU ----------- International Telecommunications Union
ITU-T -------- International Telecommunications Union-Telecommunications Sector
ITU-R -------- International Telecommunications Union - Radio/Satellite Sector
IWF ----------- Interworking Function
LAN ---------- Local Area Network
LANE -------- LAN Emulation
LI -------------- Length Indicator
LME ---------- Layer Management Entities
LOF ---------- Loss of Frame
LOP ---------- Loss of Pointer
LOS ---------- Loss of Synchronization
LTE ----------- Line Termination Equipment
MAN ---------- Metropolitan Area Network
MD ------------ Mediation Device

MID ----------- Message Identifier
MPEG ------- Motion Picture Experts Group
MUX ---------- Multiplexer
NE ------------ Network Element
NIC ----------- Network Interface Card
N-ISDN ------ Narrowband Integrated Services Digital Network
NNI ----------- Network Node Interface
OA&M ------- Operations, Administration and Maintenance
OAM&P ----- Operations, Administration, Maintenance and Provisioning
OC ------------ Optical Carrier
OSI ----------- Open System Interconnection
OS ------------ Operational System
PDU ---------- Protocol Data Unit
PHY ---------- Physical Layer
PM ------------ Performance Monitoring
PMD ---------- Physical Media Dependent
PNNI --------- Private Network Node Interface
POH ---------- Path Overhead
PRI ----------- Primary Rate Interface
PRM ---------- Protocol Reference Model
PT ------------ Payload Type
PTE ---------- Path Terminating Equipment
PVC ---------- Permanent Virtual Connection
PVP ---------- Permanent Virtual Path
QoS ---------- Quality of Service
RAI ----------- Remote Alarm Indication
RDI ----------- Remote Detection Indication
REI ----------- Remote Error Indication
RF ------------ Radio Frequency
RFI ----------- Remote Failure Indication
ROI ----------- Return On Investment
RT ------------ Remote Terminal
SAM ---------- Service Access Multiplexer
SA ------------ Service Affecting
SAR ---------- Segmentation and Reassemble
SAR-PDU --- SAR Protocol Data Unit
SAR-SDU --- SAR Service Data Unit
SDH ---------- Synchronous Digital Hierarchy
SDU ---------- Service Data Unit
SHR ---------- Self Healing Ring
SMDS -------- Switched Multimegabit Data Service
SN ------------ Sequence Number
SNMP -------- Simple Network Management Protocol
SNP ---------- Sequence Number Protection
SONET ------ Synchronous Optical Network
SPE ---------- Synchronous Payload Envelope

SS7 ----------- Signaling System 7
SSCF -------- Service Specific Coordination Function
SSCOP ------ Service Specific Connection Oriented Protocol
SSCS -------- Service Specific Convergence Sublayer
ST ------------ Segment Type
STE ---------- Section Termination Equipment
STM ---------- Synchronous Transfer Mode
STS ---------- Synchronous Transport Signal
SVC ---------- Switched Virtual Connection
SVP ---------- Switched Virtual Path
TC ------------ Transmission Convergence
TCO ---------- Total Cost of Ownership
TCP ---------- Transmission Control Protocol
TDM ---------- Time Division Multiplexing
TM ------------ Traffic Management
TMN ---------- Telecommunications Management Network
TOH ---------- Transport Overhead
TL 1 ---------- Transaction Language 1
TSA ---------- Time Slot Assignment
TSI ----------- Time Slot Interchange
TU ------------ Tributary Unit
UBR ---------- Unspecified Bit Rate
UDLC -------- Universal Digital Loop Carrier
UNI ----------- User-to-Network Interface
UPSR -------- Uni-directional Path Switched Ring
UTP ---------- Unshielded Twisted Pair
VBR ---------- Variable Bit Rate
VC ------------ Virtual Channel, Virtual Container, Virtual Connection
VCC ---------- Virtual Connection Concatenation
VCI ----------- Virtual Channel Identifier
VDT ---------- Video Dial Tone
VIP ----------- Video Information Provider
VIU ----------- Video Information User
VNI ----------- VIP-to-Network Interface
VP ------------ Virtual Path
VPI ----------- Virtual Path Identifier
VT ------------ Virtual Tributary
VTC ---------- Video Teleconferencing
VTOH -------- Virtual Tributary Overhead
WAN --------- Wide Area Network
WDM --------- Wave Division Multiplexing
W-DCS ------ Wideband DCS